•JUMPSTART•
the HP-48G/GX
featuring Engineering & Science Applications

The easy way to get started

Thomas Adams, Ph.D.
Professor of Physiology

and

Merle C. Potter, Ph.D., PE
Professor of Mechanical Engineering

Michigan State University

Great Lakes Press, Inc.
P.O. Box 483
Okemos MI 48805
www.glpbooks.com

JUMPSTART the HP-48G/GX: with Special Engineering Applications
ISBN   1-881018-18-0

© Copyright 1996 by Great Lakes Press, Inc.

All rights reserved. No part of this book may
be reproduced in any form, in whole or in part,
without prior written consent of the publisher.

All inquiries and comments should be addressed
to the Customer Service Dept:

Great Lakes Press
P.O. Box 550
Grover, MO 63040
(314)   273-6016
www.glpbooks.com

Printed in USA by Bookcrafters, Chelsea, Michigan

Table of Contents

Preface i

Introduction 1
- About the "User's Guide" 2
- The HP-48G Keyboard 3

Chapter 1- Basic Operations 7
- Turning the HP-48G ON, OFF and
  Getting out of Traps 7
- STACK Designations 8
- Deciding How Numbers are Displayed 9
- Basic Arithmetic Operations with
  Reverse Polish Notation (RPN) 10
  - Example 1-1: 11
  - Example 1-2: 11
- Basic Directory Structure 12
- The Importance of Subdirectories and
  How to Construct Them 14

Chapter 2- Using Equations 19
- Using the EQUATION LIBRARY 19
  - Example 2-1: 20
- Using the EQUATION WRITER and
  SOLVE Functions 26
  - Example 2-2: 27

Chapter 3 - Solved Problems 37
- Exercise 1: Column Loading 38
- Exercise 2: Beam Deflection 41
- Exercise 3: Electrical Resistance 42
- Exercise 4: Capacitive Energy 43
- Exercise 5: Fluid Dynamics 43
- Exercise 6: Fluid Pressure 44
- Exercise 7: Wind Force 45
- Exercise 8: Drag Force 46

Table of Contents (continued)

- Exercise 9: Isothermal Gas Expansion 47
- Exercise 10: Heat Transfer 48
- Exercise 11: Projectile Motion 50
- Exercise 12: Linear Motion 51
- Exercise 13: Terminal Velocity 52
- Exercise 14: Mass-Spring Oscillation 53
- Exercise 15: Plane Geometry 53
- Exercise 16: Solid Geometry 54
- Exercise 17: Solid State Devices 55
- Exercise 18: Shear Stress 56
- Exercise 19: Shear Stress (Mohr's Circle) 57
- Exercise 20: Roots of a Polynomial 58
- Exercise 21: Solving Equations 59
- Exercise 22: First-order Differential Equation 60
- Exercise 23: Solve a System of Linear Equations . . . 61
- Exercise 24: Second-order Differential Equation 61

Chapter 4 - Special Applications 65
- Matrix Operations 65
- Storing, Recalling and Editing a Matrix 67
- Examples of Matrix Operations 68
- Solving a System of Linear Equations 70
- Applications using Complex Numbers
  and Vectors 72
- Simultaneous Solutions for Non-linear
  Equations - An Example 73
  Background 73

Chapter 5 - Writing Programs 83
- Basics of Program Structure 83
- Using Special Characters 83
- Example for using Special Characters: 84
- Program Design 85
- Example for Symbolic Solution 86
- Detailed Program Description 91
- Keying ALPHA (α) Symbols 93
- Keying the Program 94
- Running the Program 95
- The Next Step 97

Table of Contents (continued)

Chapter 6 - Basic Statistics 101
- Example: Statistics for a Single Data Set 102
- Example: Statistics Using Array Functions 106
- Programs for Statistical Calculations 111
- The Next Step 115
- Deleting Local Variables and Subdirectories 115

Chapter 7 - A Practical Problem 121
- The Problem 121
- Background 122
  - Why Some Animals Regulate
    Body Temperature 122
  - How Body Temperature is Regulated 123
- The Physical Avenues of Heat Exchange 124
  - Heat Exchange by Thermal Conduction 124
  - Heat Exchange by Convection 125
    - Convective Heat Exchange
      Inside the Body 125
    - Convective Heat Exchange
      at The Body Surface 126
  - Heat Exchange by Thermal Radiation 127
  - Heat Loss By Evaporation 129
- Physiological Factors 130
  - Metabolic Heat Production 130
- Human Variations in Heat Strain 131
- The Human "Heat Stress Index" 132
  - General Description 132
  - Using the program 133
  - Symbol Definitions 135
  - Equations 135
- Sample Problems 136
  - Program Listing 141

Chapter 8 - Procedures for File Transfer 147
- Transferring Files between HP-48G's 148
  - Generating the Subdirectories and
    Their Files 149

Table of Contents (continued)

- Transferring One or More
  Subdirectories 150
  Transferring One or More
  LOCAL VARIABLES 152
- Transfers Between the HP-48G and the
  HP-48S, and *vice versa* 153
  File Transfer from an HP-48S to an HP-48G 153
  File Transfer from an HP-48G to an HP-48S 154
- Saving the HP-48G View Window
  Display to a Disk 156
- File Transfer from an HP-48G to a
  Desktop Computer 158

Index **165**

Preface

This book has only one goal - to describe for a beginner the basic features and functions of the HP-48 (G and GX models) handheld calculator/computer. It takes readers from the initial confusion most have when they first take the machine out of the box and try to use it, through a process by which they eventually master this marvelous machine. This book does not presume to describe all there is to know. That comes later from reading more advanced books, self-study and lots of practice. But, this book does something more important! It gets you started.

It's hard to overstate the HP-48's value for those in engineering, medicine, the basic sciences or other technical fields. Among its many features, it helps with all basic mathematical procedures and converts physical units with ease. It also does statistics and curve-fitting operations, plots graphs, solves equations, and allows writing even complex programs with many decision statements, loops and flags. It can be linked to a desktop computer and used to drive a printer or other devices. For the money invested, no other technology returns such power. There are lots of graphing calculators and handheld computers around these days, but the HP-48 is unarguably the top of the line for the price.

That's only some of the good news. Perhaps most important, the HP-48 is surprisingly easy to use. It is difficult, though, for most people to get started learning to use it. That's as true for an engineering freshman, as it is for the newly graduated senior, as it is for the Ph.D.-level user with many years of professional experience. This difficulty is not because there's something wrong either with the machine's design, or with the people trying to use it. Quite the contrary. The HP-48 is expertly configured and crafted, and most users have all the basic skills to take full advantage of it. The rub comes from something else. Few are practiced experts with this technology. We all have to start at the beginning. But, it doesn't take more than just a few hours to learn the basics and to start using the full power of this remarkable instrument.

Why is the HP-48 so important? It doesn't present any mathematical procedures that haven't been around for a long time.

Preface

It doesn't draw any graphs that can't be done with paper-and-pencil. It doesn't make any calculations that can't be done in any number of other ways. So, what does it do that's so special? Similar questions can be asked of any tool. What's so important about a car? I can always walk to the next town. What's so important about a screwdriver? I've put screws in with a dime before, and I can do it again.

The HP-48 does what every other successful tool does - it makes the job faster, easier and provides free time for more useful human activities than just the task itself. In technical and professional areas, the HP-48 focuses attention on more intellectually appropriate activities than just making calculations. It facilitates problem solving.

This book starts the process of learning how to go quickly and accurately from one computational place to another, and how to get what you want from the HP-48. It also promotes going beyond basic skills to learn more about mathematical procedures. Most important, it provides a tool for developing professional and technical insight through problem solving. A case can be made that the HP-48 would be useful for teaching math and physics like this in the first place. Few of us had that opportunity, but fortunately, it's never too late. Start reading at page 1. Go slowly, don't overlook anything and take it a step-at-a-time. The return will be enormous.

Thomas Adams, Ph.D.
Department of Physiology
Michigan State University

E-mail: adamst@pilot.msu.edu

Merle C. Potter, Ph.D., P.E.
College of Engineering
Michigan State University

E-mail: MerleCP@aol.com

Introduction

The HP-48G and the HP-48GX are the newest models in a long line of progressively more sophisticated calculators from Hewlett-Packard. They are powerful and with a little practice, easy to use. They have many built-in features for solving equations, converting among physical units, writing programs, generating graphics, producing print-outs, and dealing with many other operations. Also, data and programs are easily shared between calculators by infrared transmission, and just as readily stored in the hard drive of a desktop computer, or on standard 3.5" disks using an interface cable. Exploring the many features of this relatively inexpensive instrument is as interesting as it is valuable in learning how to use it to perform even sophisticated mathematical operations.

All instruction in this book is applicable to both the HP-48G and the HP-48GX. Both will be referred to as the "HP-48G". Descriptions here present only a few of the HP-48G's many features. They will show, for example, how to obtain solutions with the machine's extensive store of equations in its EQUATION LIBRARY. They will also show how to use the HP48's EQUATION WRITER facility to construct and solve equations either numerically or symbolically that are not held in its own library. This opens an important door of opportunity for the user to apply this instrument's power to any number of unique applications.

This book also describes step-by-step procedures for writing, storing and using programs for the HP-48G. It illustrates how to construct alphabetic and numerical values for data prompts, generate related local variables, write applicable equations, then display answers with customized statements and appropriate units. All the IF/THEN, IF/THEN/ELSE, DO/UNTIL, FOR/NEXT/STEP, CASE/THEN, WHILE/REPEAT, full use of Flags and other logic controls are available to the HP-48G user. Thanks to the machine's permanent memory, all programs and stored data are held in RAM when the calculator is turned off, and even when its batteries are being replaced.

Like so many HP-48G's applications, there are several different, equally valid and accurate ways of arriving at solutions. It is up to the user to decide which one is best in a particular application. The first step in this process, though, is to become familiar enough with the instrument to be able to use it with some degree of ease. There's just no getting around having to know what each key does and what are the built-in features of the calculator. It is the same for operating any other intricate instrument. For most people, this may take some effort at first, but the process soon becomes much easier and the rewards progressively more valuable.

To get it all started, Chapter 1 gives only the bare-bones of basic operations of the HP-48G. It describes important definitions and uses for the STACK REGISTERS, and shows how to customize display features. It also reveals all the not-so-mysterious processes involved in "Reverse Polish Notation" and demonstrates with a couple of examples how simple and powerful it is. This chapter also shows how to construct subdirectories and why this is an important technique. These sections provide the basic tools for interested users to continue exploration of the calculator on their own. They also set the stage for using the HP48's EQUATION LIBRARY, EQUATION WRITER, for developing mathematical models, for programming and for reading more advanced and comprehensive guides.

About the "User's Guide"

Each new HP-48G comes with a comprehensive instruction manual called the "User's Guide". This is an excellent source of detailed information. Its encyclopedic quality, however, makes it more useful as a reference, than it is as an instruction manual. Although expertly organized, complete and precisely written, it assumes a level of insight into programming, statistical and mathematical technologies that is not appropriate for all users. Most people will benefit from first reading introductory information about the HP-48G, like that provided in this book.

Introduction

The HP-48G Keyboard

The keyboard of this marvelous instrument may be somewhat bewildering at first. Its complexity bespeaks the many services it can deliver. Nevertheless, knowing what operation each key performs is an essential first-step. Even before reading any further, it would be well worth the few minutes it takes to examine the keyboard carefully, just to see the different names on and around each key. The point is not to decipher what each does, but to gain general familiarization with where everything is.

The most obvious functions of the keyboard are those performed by the keys of the number pad and the keys at the lower right of the keyboard that control arithmetic operations. The sequences of keystrokes required to complete arithmetic solutions using the "Reverse Polish Notation" (RPN) logic of the HP-48G are explained in the first chapter.

Many keys are marked to show they control up to four different operations, each of which is designated on or near the key. The "primary function" of a key, for example is shown in white letters on its black face. The "left-shifted function" (LS) of a key is shown in purple, typically at the upper left or at the top of the key. The "right-shifted function" (RS) of a key is shown in green, typically at the upper right or at the top of the key. There are six, white-faced keys at the top of the keyboard designated "A" to "F". They are used to activate menus and variables that will appear in corresponding locations at the bottom of the view window when the calculator is turned on. A little familiarization soon shows each key does a lot more than control just the operations for which it is marked.

In this book, the "left-shifted function" of a key is designated by "LS". Its "right-shifted function" is designated by "RS". "RC" means "right cursor" control (▶) ; "LC" means "left cursor" control (◀).

Introduction

Many keys are also used to generate alpha (α) characters. This function is controlled by first pressing the key with the α symbol just under the ENTER key, then pressing the key appropriate for the required letter, as shown at its lower right. Instructions later in the book show how to generate alphanumeric strings, along with upper and lowercase sequences, and a wide range of "special characters", like δ, β, π, ≤, Σ, Å, among many others.

Congratulations! You have started an exciting adventure.

> This book gives just the basics for operating the HP-48G and HP-48GX. For more complete instruction see: *"Mastering the HP-48G/GX"*, Thomas Adams, Kendall Hunt Publ. Co. (1-800-228-0810)

Chapter 1

Basic Operations

- Turning the HP-48G On and OFF and Getting Out of Traps

- STACK Designations

- Deciding How Numbers are Displayed

- Basic Arithmetic Operations with Reverse Polish Notation (RPN)

- Basic Directory Structure

- The Importance of Subdirectories and How to Create Them

Notes

Chapter 1- Basic Operations

Turning the HP-48G ON, OFF and Getting out of Traps

The beginner isn't going to get very far without knowing how to turn the HP-48G on, off and how to get out of traps during operations. An improperly designated function, or an incorrectly keyed entry can sometimes cause the calculator to appear locked. No key is active, the display screen remains blank or unchanged in its image, and the frustrated user is confronted only with an error-indicating BEEP, no matter what operation is attempted.

Most often, the computer is not locked, but it's waiting for the next legitimate keystroke or data entry, but one unknown to the user, and for which an explicit request is not generated. All too often, not even turning the HP-48G off, or taking out its batteries, is enough to provide recovery. What a predicament! It's easily solved, as described in the next paragraph.

Turning the HP-48G on is simple enough - just press once the key at the lower left corner of the keyboard whose primary function is "ON". Turning it off is just as easy. Activate the right-shifted function of the same key by first pressing RS, then pressing the ON key. Unique among other keys, the ON key has an additional function designated immediately below it. No matter what operation is being performed, pressing the ON key once or twice when the instrument is activated (on), cancels the operation. This is the way to get out of traps. This function may not clear the view window, but the keyboard will become active again while the previous operation is abandoned.

> When the HP-48G is on, pressing the ON key activates a CANCEL function that abandons whatever operation is currently active.

STACK Designations

To activate accurately some functions directly from the keyboard requires knowing how lines in the display window are named. Although each obviously has a number, each also has a STACK designation. Unfortunately, this important information is neither available from the display, nor from information on the keyboard itself. Knowing their code, however, is essential.

When the HP-48G is first turned on, lines in the view window are numbered 1 to 4. Line 1 is designated as the "X STACK REGISTER", line 2 is the "Y STACK REGISTER" (Figure 1-1). Line 3 is the "Z STACK REGISTER" and line 4 is the "T STACK REGISTER". The last two names are less important for accurate keyboard operations.

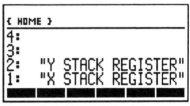

Figure 1-1: X and Y Registers

It is essential to remember that the lowest level of the STACK is a register called "X" and the next one up is called "Y". For example, the transcendental functions (SIN, COS, TAN, ASIN, ACOS and ATAN) operate only on whatever number is in the X STACK REGISTER (Line 1 of the view window). Keying "1234", then SIN yields "0.4384" (information in the next few sections describes how to set the digit display). Similarly, pressing the "1/x" key when "4,321" is at Line 1 calculates "2.314E-4" - the calculation was performed on the number in the X STACK REGISTER. The answer format depends on display configuration.

Other keyboard operations also require knowing about the X and Y STACK REGISTERS. For example, the "y^x" key raises the number in the Y STACK REGISTER to the power designated by the number in the X STACK REGISTER. Keying "1.23", then pressing ENTER places that number in the X STACK REGISTER. Next, keying "45.6", then pressing ENTER, raises "1.23" to the Y STACK REGISTER and places "45.6" in the X STACK REGISTER. Pressing "y^x" yields "12,579.78".

The use of the symbol "X" to designate the lowest register in the view window has no relationship either to the apparently similar symbol generated as the α definition of the reciprocal key (1/x), or to the "X" symbol on the key at the lower right of the keyboard that controls the multiplication function. Although these symbols resemble one another superficially, they control very different operations. The "X" symbol, however, of the "$\sqrt{x}$", "y^x", "1/x" primary key functions and the shifted functions "x^2", "$\sqrt[x]{y}$", "10^x", and "e^x", all refer to the number in the X STACK REGISTER.

Deciding How Numbers are Displayed

There is a wide range of options for displaying numbers. No matter what style is used, the precision of a calculation is unaffected - it retains twelve digits to the right of the decimal.

How numbers are displayed does not affect the precision of a calculation.

Numbers displayed in a STANDARD notation show all significant digits. Those shown in a FIX mode allow the user to select the number of digits to the right of the decimal. Those in SCI or ENG also allow designating the number of digits shown to the right of the decimal, but use exponential notation. SCI increments the exponent in units; ENG increments it by 10^3.

The easiest way to select a format style for numbers is to select among the options in the CALCULATOR MODES menu (Figure 1-2). They are available with the right shifted function of the CST key. The initial menu is highlighted at "Std". The display in Figure 1-2

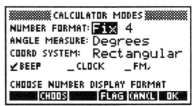

Figure 1-2: RS Modes

shows this has been changed to "Fixed". To select one of the other display options, press the white key at the top of the key board marked "B" directly under the "CHOOS" function displayed at the bottom of the view window. This presents a new menu that lists the different styles for displaying numbers (Figure 1-3).

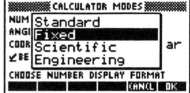

Figure 1-3: Format Menu

Use the up (▲) or down (▼) cursor control keys to select the preferred style, then select "OK" by pressing the white key marked "F". The number that indicates the positions to the right of the decimal is shown next to "Fix". Move the cursor to the right, then key a number and press ENTER.

The CALCULATOR MODES menu (Figure 1-2) is also used to control whether angle measures are "Degrees", Radians", or "Grads", whether the coordinate system is "Rectangular", "Polar", or "Spherical", whether the tone generator ("BEEP") is activated, and to select other operating features for the machine. Each category of choices for a characteristic is viewed by first moving the cursor either vertically or horizontally with one of the control keys, highlighting the characteristic, then selecting "CHOOS". When selected configuration choices are listed in the CALCULATOR MODES menu, keying "OK" displays the HOME menu again.

The selected styles for displaying numbers and how the instrument is configured remain active indefinitely for mathematical, statistical and equation-solving operations, and will be retained when the instrument is turned off. In most applications, it can be changed anytime right from the keyboard by reactivating the CALCULATOR MODES menu and designating a new set of choices. Also, each of these features can be changed selectively in the main body of customized program.

Basic Arithmetic Operations with Reverse Polish Notation (RPN)

Using RPN makes even complicated calculations easy, either with manual operations right from the keyboard, or with those written into a program. Elements in an equation are operated on in pairs, with

Chapter 1 - Basic Operations

one in the X STACK REGISTER and the other in the Y STACK REGISTER. If necessary, the contents of these registers can be reversed by keying LS, SWAP. Repeating the keystroke sequence, LS, SWAP, toggles them to their original positions. For example, key "1", ENTER, "2", ENTER, then LS, SWAP a couple of times. LS, CLEAR, clears the STACK.

A couple of examples will show the process of RPN to be direct and straightforward. It obviates using parentheses or brackets in an equation solution, takes less time than other methods and readily provides interim results with each arithmetic operation.

Example 1-1: What are the keystrokes required to solve the following equation?

$$X = \frac{(30)(6) - (10+17)}{5}$$

Solution: (Hint: the symbol "$*$" indicates multiplication)

30, ENTER, 6, $*$, 10, ENTER, 17, +, -, 5, ÷ (read: "30.60")

Example 1-2: What are the keystrokes required to solve the following equation?

$$X = \frac{(\frac{31.2}{9.6})^{1.32} (LOG\ 4.9 + (3.2)^2)}{(2.01)^3}$$

Solution:

31.2, ENTER, 9.6, ÷, 1.32, y^x, 4.9, ENTER, RS, LOG, 3.2, ENTER, LS, x^2, +, $*$, 2.01, ENTER, 3, y^x, ÷ (read: "6.38")

These simple exercises used no more than Lines 1, 2 and 3 of the view window for data storage. Many more lines, up to memory limits of the calculator, could be used if necessary for longer sequences of data entry. Although they would extend off-screen, they could be reviewed and edited using the ▲ or ▼ cursor controls.

Despite the amount of data storage in the STACK, each arithmetic operation involved only data in the X and Y STACK REGISTERS. If this wasn't obvious when the calculations were made for the first time, it'd be a good idea to repeat them and watch the view window as each step is activated. Exercises later in this book show how both alphabetic and numerical data in the STACK can be duplicated, erased, or otherwise manipulated and rearranged. These processes are done directly from the keyboard, or through instructions in a program.

Basic Directory Structure

The six, blue rectangles at the bottom of the view window represent locations at which variables are stored (Figure 1-4). They could be numbers, entire programs, special characters, alphanumeric strings, or other objects.

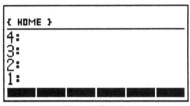

Figure 1-4: The HOME Directory

When more than six such objects are stored in a directory, a new "page" is automatically constructed to accommodate the requirement for the extra memory locations. Sometimes variables are constructed and stored as a routine operation in using the EQUATION LIBRARY, the EQUATION WRITER, or in the operation of a customized program. The number of "pages" is expandable to the memory limit of the machine. Press NXT to see additional "pages".

For practical applications, it is cumbersome to have to navigate multiple "pages" to locate a required variable. Fortunately, there are alternatives, as described in the next couple of sections, to purge variables either manually or automatically when they have outlived their usefulness. Also, an alternative is to construct series of subdirectories. Their importance is greater than just limiting the number of "pages" of stored variables, as described in the next section.

Chapter 1 - Basic Operations   13

A simple example shows how variables are constructed from the keyboard, how they are used, and how they are eliminated when they are no longer needed. The numbers "1" to "7" will be stored in variables called "A" to "G", respectively, in the HOME directory.

First, key "1", then press ENTER. Next, press the apostrophe key (1st key in the 2nd row of black keys), then press in sequence the keys α, A, STO. This procedure stored the number "1" in a local variable called "A" which was then displayed at the bottom of the view window on the left.

To recall the number from storage, press the white key marked "A" that coincidentally corresponds to the variable of the same name. The number is called from storage and copied to the X STACK REGISTER. A second copy is easily made by again pressing the "A" key.

Removing a number from Line 1 requires pressing the erase key (5th key, 4th row; the left pointing arrow). Pressing it again clears the view window. Repeat the sequence for generating variables to store the numbers from "2" to "7" as variables "B" to "G", respectively. Key: 2, ENTER, ', α, B, STO, and so on.

The stored variable "A" disappeared from the listing at the bottom of the view window when the one called "G" was created. "A" remains in memory, but the menu location controlling it has been moved to the next "page". Pressing NXT displays the next "page". Pressing LS, PREV redisplays the previous "page". Regardless of what page is in current display, pressing a white key corresponding to a variable copies its contents to the STACK.

Typically, the user-constructed variable menu, like the one produced in this exercise, will be replaced by others specific to HP-48G built-in applications. For example, RS, UNITS displays a series of "pages", each of which designates subdirectories appropriate for performing unit conversions. Pressing VAR recalls the list of stored variables for the exercise. VAR performs this function regardless of what is currently in display.

To recall the contents of all the variables for this example, press the white buttons for each menu location for both pages. When the last variables have been recalled, the first ones that appeared are now displaced off-screen to the top of the view window. Although out-of-view, they can be displayed by using the ▲ cursor control key. LS, CLEAR erases all STACK entries, including those that are not displayed.

It is easy to remove the variables created in this exercise. First key LS, { }, then press each of the white keys for both pages. When all variables are displayed between the "{ }" symbols, press ENTER, then LS, PURG. The HOME directory is again clear of all stored variables.

The Importance of Subdirectories and How to Construct Them

When a new HP-48G is first turned on, the view window shows at its upper left that the machine is now in the "HOME" directory (Figure 1-5). The six blank menus at the bottom of the screen show there are no data, programs, or other objects defined as variables.

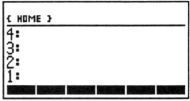

Figure 1-5: The HOME Directory

It is a good idea to construct subdirectories within the HOME directory for each program, equation

> Create an appropriately named, but new, subdirectory for each problem, equation solution, use of the EQUATION LIBRARY, program or other application.

solution, statistical problem, or other application, including the use of the HP-48G's EQUATION LIBRARY. Not only does this reduce the number of "pages" required to store local variables, it also allows identically named variables to be used independently in each application and in each subdirectory.

As an example: to create a subdirectory called "SUB1", key RS,

MEMORY, then press the white key "D" for "NEW" to display the "NEW VARIABLE" menu (Figure 1-6). Press the ▼ key once, then key α, α, S, U, B, 1, α, ENTER, press the white key "C" to select "✔CHK", then press the white key "F" for OK.

Figure 1-6: NEW VARIABLE Menu

The "OBJECTS IN {HOME}" menu now lists the subdirectory called "SUB1" (Figure 1-7). Press CANCEL to see the subdirectory listed at the far left as a variable selection. Pressing the white key "A" to select it displays there are no variables listed for the subdirectory. Notation at the top left of the view window indicates that the calculator is now operating in a named subdirectory (SUB1) of the HOME directory (Figure 1-8).

Figure 1-7: Objects in HOME

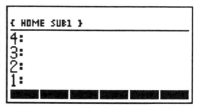
Figure 1-8: SUB1 subdirectory

Moving from the HOME directory to any subdirectory only requires pressing the white key corresponding to the view window listing for the subdirectory. A small tab at the left of a variable location identifies it as a subdirectory (Figure 1-9). Moving from a subdirectory back to the HOME directory requires keying either RS, HOME, or LS, UP if there is only one level of subdirectory structure, as there is in this example of "SUB1".

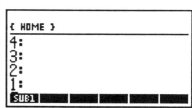
Figure 1-9: Inside SUB1

To purge a subdirectory and all its variables, key LS, MEMORY, then press "DIR", press the apostrophe key, α, α, key (name of subdirectory), α, ENTER, then press "PGDIR". This operation must be done only with care and consideration, because not only is the subdirectory removed from memory, so are all its variables. None can be recalled.

This chapter presents important basic features and operations of the HP-48G. Knowing about them is fundamental to even the simplest applications and calculations. Repeating the exercises in this chapter enough times that all these processes and styles are second-nature gives a good foundation for what comes next.

The next chapter introduces the HP-48G's EQUATION LIBRARY and shows how to make some sample calculations with it.

Chapter 2

Using Equations

- The EQUATION LIBRARY

- Using the EQUATION WRITER and SOLVE Functions

Notes

Chapter 2- Using Equations

It's not hard to predict that HP-48G owners will want to use their machines to solve equations, one way or the other. It was, of course, designed to meet these needs.

This chapter introduces two powerful ways to solve equations. The EQUATION LIBRARY provides access to a wide range of standard equations in many different areas. The EQUATION WRITER allows the user to write and solve equations of custom design. The same numerical example is used in this chapter to demonstrate similarities for using these two techniques.

Using the EQUATION LIBRARY

The EQUATION LIBRARY is a sizeable listing of equations in ROM commonly used in engineering and physics (Figures 2-1 to 2-3). They cover calculations for "Columns and Beams", "Electricity", "Fluids", "Energy", "Gases" and many other areas.

Figure 2-1:
Equation Library

Figure 2-2:
Equation Library

Figure 2-3:
Equation Library

Their utility is, of course, that the user doesn't have to

Chapter 2 - Using Equations

generate equations or programs, but more readily, just accesses those already in memory. This doesn't preclude, however, writing equations of one's own design, or constructing a program to solve one or more equations, as described later in this chapter.

Regardless of the category in which equations are to be solved, the procedures for using the EQUATION LIBRARY are all about the same. Solving a sample problem with these procedures shows how easy it is.

Example 2-1:

Calculate the thermal conductivity of a 6 inch thick wall with a surface area of 200 square feet that has an inside temperature of 75°F and an outside temperature of 32°F across which there is steady state heat flow of 700 Btu per hour.

Solution:

Step	Operation
1	Create a subdirectory called DEMO, then enter that sub-directory by pressing the white key marked "A" while in the HOME directory (Figure 2-4, next page).
2	The solution to this problem will be made using the HP-48G EQUATION LIBRARY. Key RS, EQ LIB to generate the display shown in Figure 2-5, next page). Because data will be entered using English units, select ENGL by pressing the white key marked "B". Because

Chapter 2 - Using Equations   21

units are required in displays, select UNITS by pressing the white key marked "C". A small white square appears when each choice is correctly set. The down-pointing arrow in the lower right corner of the view window signifies there are other options off-screen, but the one required for this solution is at the last line of the display.

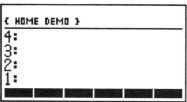
Figure 2-4: DEMO subdirectory

Figure 2-5: RS, EQ LIB

3   Highlight the HEAT TRANSFER option by using the down-cursor control key (▼) to generate the display shown in Figure 2-6, then press ENTER to see the display in Figure 2-7.

Figure 2-6:
Select "Heat Transfer"

Figure 2-7:
"Heat Transfer" menu

Chapter 2 - Using Equations

4 Use the ▼ key to highlight "Conduction" (Figure 2-8), then press ENTER to see the first of two equations in the display, as shown in Figure 2-9.

Figure 2-8:
Highlight "Conduction"

Figure 2-9:
See first equation

5 To see a graphic that helps visualize the problem, press the white key marked "D" to select the menu item at the bottom of the view window designated PIC (Figure 2-10). To see a set of definitions for the variables associated with this solution, press the white key marked "C" to select VARS (Figure 2-11). Some are off-screen, but can be read by using ▼.

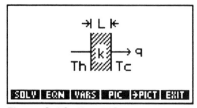

Figure 2-10: Key PIC

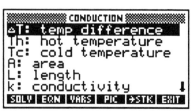

Figure 2-11: Key VARS

Chapter 2 - Using Equations

6   Press the white button "B" to select EQN to see the first equation again (Figure 2-12). Press NXEQ to see the second equation (Figure 2-13) which is more appropriate for this problem.

Figure 2-12: Key EQN

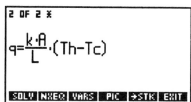
Figure 2-13: Key NXEQ

7   Press the white key marked "A" to select SOLV and begin the solution of the problem (Figure 2-14; next page). The menu items at the bottom of the display designate the local variables into which numerical data will be stored. To start the process, use the number pad to key "75", then press the white key marked "B" to store this value as the variable "TH" (Figure 2-15; next page). The value with appropriate units is displayed at the upper left of the view window. Enter data for "TC", "A" and "L" in a similar manner. Do not enter data for "K", of course, because this is the variable for which a solution will be made.

24 Chapter 2 - Using Equations

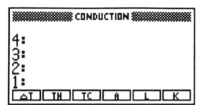

Figure 2-14: Key SOLV

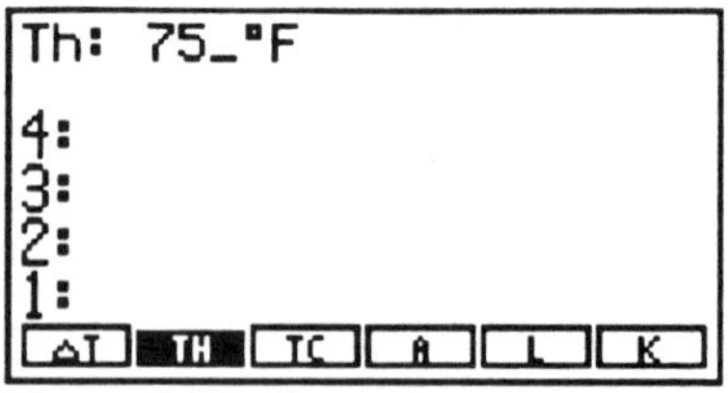

Figure 2-15: Enter data for "Th"

8   Press NXT to go to the next "page" to enter data for "Q" (Figure 2-16). Press LS, PREV to return to the previous "page". To solve for "K", press LS, then press the white key marked "F" to select the variable "K" for which a solution is required (Figure 2-17).

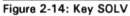

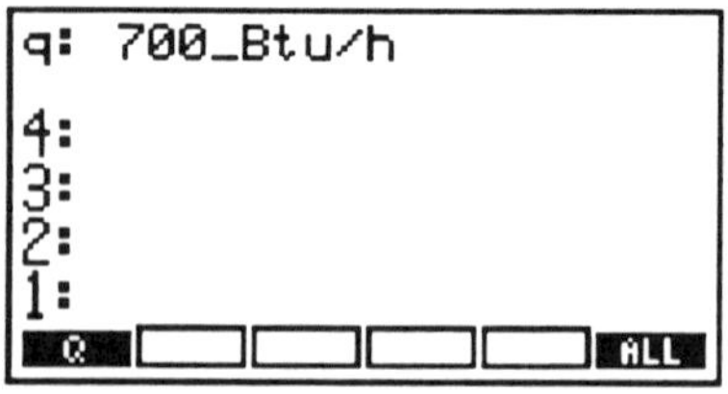

Figure 2-16:
Key NXT then enter data for "Q"

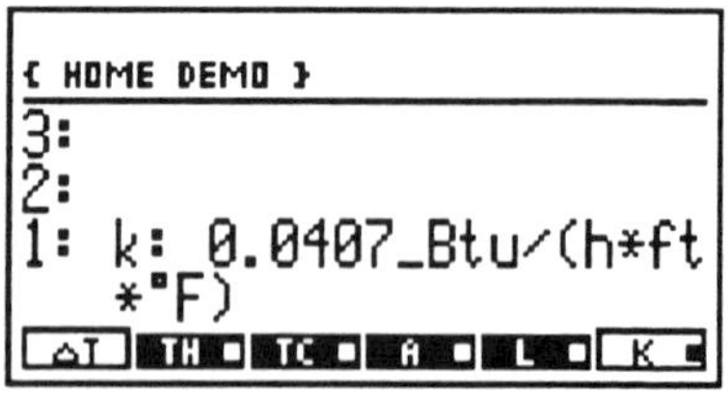

Figure 2-17:
Key LS, K to obtain solution

Chapter 2 - Using Equations   25

9   Key VAR to review local variables in the DEMO subdirectory (Figure 2-18) with the solution for "k" still displayed at line 1 of the view window. Review the numerical values for "Q", "L" and "A" by pressing the corresponding white keys (Figure 2-19).

Figure 2-18: Key VAR   Figure 2-19: Key Q, L and A

10 To remove the local variables from this subdirectory, first press LS, { }, then press the appropriate white keys to select "EQ", "MPAR", "Q", "K", "L", and "A", press NXT, select "TC", "TH", and "∆T". Press ENTER (Figure 2-20, next page), then key LS, PURG. The view window now shows that the subdirectory DEMO contains no local variables (Figure 2-21, next page).

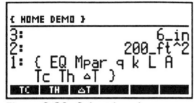

Figure 2-20: Select Local Variables

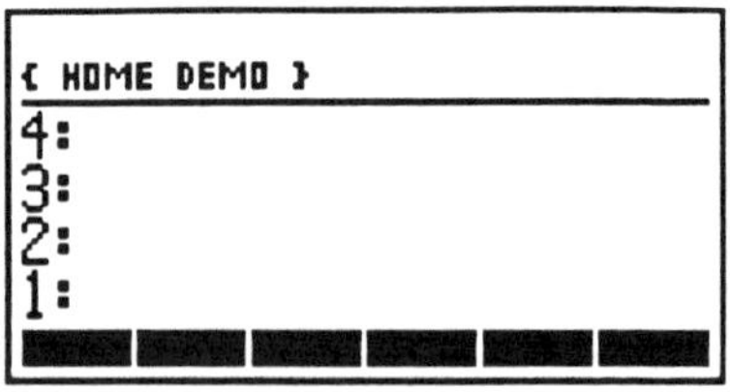

Figure 2-21: LS, PURG

> *The EQUATION WRITER and SOLVE functions support solving any equation, not just those in the EQUATION LIBRARY.*

Using the EQUATION WRITER and SOLVE Functions

The EQUATION WRITER and SOLVE functions of the HP-48G used together give valuable options for solving virtually any equation. They are practical alternatives to using the EQUATION LIBRARY, and easily allow for solutions of equations that aren't listed in it. Using them is a three-step process. First, an equation is written in the view window using the EQUATION WRITER. This feature is activated as the left-shifted function of the ENTER key. Second, the equation is stored as a local variable using an easily-remembered, user-defined name. Finally, the right-shifted function of the "7" key (SOLVE) begins the process of entering numerical data and obtaining a solution. A short exercise shows how straightforward the process is.

Chapter 2 - Using Equations

For comparison, the same problem used to demonstrate solutions using the EQUATION LIBRARY is used again to demonstrate combined uses of EQUATION WRITER and SOLVE.

> Hint: Purge all local variables in a subdirectory before beginning a new problem.

Example 2-2:

Calculate the thermal conductivity of a 6 inch (0.5 foot) thick wall with a surface area of 200 square feet that has an inside temperature of 75°F and an outside temperature of 32°F across which there is steady state heat flow of 700 Btu per hour.

Solution:

Step	Operation
1	The solution will be made in a subdirectory called "DEMO" from which all previously generated local variables have been purged (Figure 2-22, next page). Purge variables for each new problem. Keying LS, EQUATION gives the display in Figure 2-23, next page.

28   Chapter 2 - Using Equations

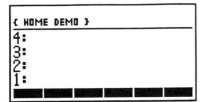

Figure 2-22:
The DEMO subdirectory

Figure 2-23:
The EQUATION WRITER

2   The equation is written into the EQUATION WRITER view window following a few simple steps. The following procedure is by no means the only or simplest way to construct the equation for this example. Steps are chosen to show some alternatives for constructing symbols, upper and lowercase characters and other features.

Key: α, LS, Q, LS, =,LS, ( ), α, LS, K, *, α, A, RC, ÷, α, L, RC, LS, ( ), α, α, T, LS, H, -, T, LS, C, α, RC

This sequence writes the equation, as shown in Figure 2-24, next page. Pressing ENTER, exits the EQUATION WRITER facility and places the equation in the X STACK REGISTER at Line 1 of the view window (Figure 2-25, next page)

$$q = \frac{(k \cdot A)}{L} \cdot (Th - Tc) \square$$

Figure 2-24: Write the equation

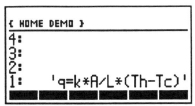

Figure 2-25: Equation at Line 1

3   The next step is to store the equation with an easily remembered name. Press the apostrophe key, then α, α, H, E, A, T, α, ENTER, to generate the display shown in Figure 2-26. Press STO to store the equation in a local variable arbitrarily called "HEAT" (Figure 2-27).

```
{ HOME DEMO }
4:
3:
2:    'q=k*A/L*(Th-Tc)'
1:                'HEAT'
```
Figure 2-26: Designate a name

```
{ HOME DEMO }
4:
3:
2:
1:
[HEAT]
```
Figure 2-27: Store the equation

4   The first step in solving the equation is to key: RS, SOLVE (Figure 2-28, next page), then with "Solve equation" highlighted, press

Chapter 2 - Using Equations

"OK" to show the SOLVE EQUATION menu (Figure 2-29).

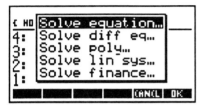
Figure 2-28:
Highlight "Solve Equation"

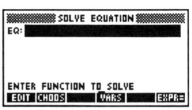
Figure 2-29:
The SOLVE EQUATION menu

5 Press CHOOS (Figure 2-30). Were other equations stored in this subdirectory, they would be listed in subsequent lines. The one selected for solution would first be highlighted by using the ▼ or ▲ cursor control keys, then pressing OK to display a data list (Figure 2-31). Because HEAT is currently the only equation listed, it is automatically highlighted as soon as CHOOS is selected.

Figure 2-30: Find the equation

Figure 2-31: The data list

6   Enter data into highlighted locations by first keying the appropriate number (Figure 2-32), then pressing ENTER. For "Q" in this example, this is "700". The cursor automatically moves to the next location for data entry (Figure 2-33). No data are entered for "k", of course, since this is the unknown variable.

Figure 2-32:
Key data for "Q"

Figure 2-33:
Press ENTER to list data

7   Use the cursor control keys to highlight other sites for data entry, key the correct number for each, then press ENTER for each (Figure 2-34, next page). For consistency of units, a value for "L" is entered as "0.5" (feet). Return the cursor to highlight the site for the unknown variable, then press the white key "F" for SOLVE (Figure 2-35, next page).

Figure 2-34: Enter remaining data Figure 2-35: Key SOLVE for "k"

8   To see the full expression for "k", press EDIT (Figure 2-36). To exit the SOLVE operation, press the white key "E" to select CANCL, then press NXT, and finally CANCL (Figure 2-37). The value for "k" is at Line 1.

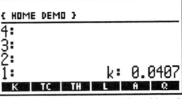

Figure 2-36: The value for "k" Figure 2-37: Value for "k" at Line 1

9   To review data for the solution, press, for example, TC, TH, L, and A (Figure 2-38, next page). The value for "K" remains in Line 5 of the STACK, but is now off-screen. To clear

Chapter 2 - Using Equations   33

the STACK, key: LS, CLEAR. To eliminate the local variables generated in this subdirectory, key: LS, { }, K, TC, TH, L, A, Q, the NXT to go to the next "page", then finally EQ and HEAT. Next, press ENTER (Figure 2-39). Keying LS PURG erases the designated local variables and LS, CLEAR returns a blank screen and an empty subdirectory.

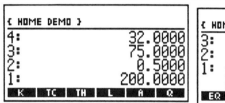

Figure 2-38: Review Data   Figure 2-39: Variables to be purged

The same as for all the other features of the HP-48G, a little practice goes a long way in making the EQUATION WRITER and SOLVE operations useful tools. Although most people will have a little trouble at first finding the right key to press and discovering what is the next best step in problem solving, procedures soon become second-nature with experience. When they do, then the HP-48G becomes an even more valuable tool. An intriguing bonus in using the HP-48G is that it is such a powerful instrument, there is always something new to learn and there is always some new feature to master.

The next chapter gives examples for solved equations.

Notes

Chapter 3

Solved Problems

- Columns and Beams
- Electricity
- Fluids
- Forces and Energy
- Gases
- Heat Transfer
- Motion
- Oscillations
- Plane Geometry
- Solid Geometry
- Solid State Devices
- Stress Analysis
- Roots of a Polynomial
- First-order Differential Equation
- Matrix Equation (a system of linear equations)
- Second-order Differential Equation

Notes

Chapter 3 - Solved Problems

In the same way that a picture is worth a thousand words, a demonstration is worth at least that many explanations. All the words are sometimes not as effective as just seeing it. Don't believe it? Write down all the step-by-step operations required to hammer a nail:

> "Grasp the middle of a short, small diameter iron rod, that is sharp at one end and flat at the other, between the thumb and the digital pad of the first finger of the non-dominant hand. Place the sharp end of the rod on the material into which it will be inserted with the rod's shaft perpendicular to the surface. Using the dominant hand, firmly grasp any solid mass of comfortable weight and using the extensor muscles of the shoulder and upper arm, raise it to be centered over the flat end of the rod...(etc.)"

It'd take ten pages to describe it all, but only ten seconds to show it, with the demonstration being remembered much better. The same principle applies to mathematical calculations and to many other techniques.

This chapter is designed to demonstrate sample calculations using the HP-48G. Although all exercises are not of equal practical use to everyone, working through them is important to all. They demonstrate generally applied techniques and will improve familiarity with the machine's many features. It's a good bet that anyone would learn something from keying each exercise. Before starting an exercise, create or use a subdirectory for it. The last section in Chapter 1 describes how to create and use subdirectories.

Use a subdirectory for the following exercises

Exercise 1: Column Loading

Problem Statement:

A vertical steel beam is fixed at its bottom and is free at its upper end. Use data in Table 3-1 to calculate the critical load (Pcr; kN) that can be applied at its top.

Table 3-1: Beam Data				
Beam	Factor	Symbol	SI units	Eng. units
10	diameter	D	cm	in
10	length	L	m	ft
2	eff. length factor	K	(none)	(none)
5	radius of gyration	r	cm	in
2.1×10^8	modul. of elasticity	E	kPa	psi
1.963×10^7	moment of inertia	I	mm^4	in^4
78.5	area	A	cm^2	in^2
notes: $r = D/2$; $I = \pi r^4/4$; $A = \pi r^2$				

Solution (SI units)

This exercise, and all others in this chapter, are completed in a subdirectory called, "EXER" for "exercises". It is created by: RS, MEMORY, NEW, ▼, α, α, E, X, E, R, ENTER, ✓CHK, OK, CANCEL. Pressing the white key "A" corresponding to "EXER" enters the subdirectory. (Reminder: "LS" and "RS" indicate "left-shift" and "right shift", respectively.)

Key the following steps:

    1. RS, EQ LIB (see Figure 3-1; next page)
    2. ("Columns and Beams" is highlighted), ENTER (see Figure 3-2, next page)

Figure 3-1

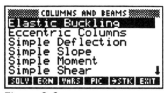
Figure 3-2

3. ("Elastic Buckling" is highlighted), ENTER (equation to calculate "Pcr" is displayed, Figure 3-3. Key EQN to see equation in EQUATION WRITER format (Figure 3-4). Key NXEQ (white key "B") three times to see other related equations in the same format, then key NXEQ, ENTER to return to display shown in Figure 3-3.

Figure 3-3

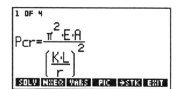
Figure 3-4

4. Key PIC to see model for problem at far right of view window for which "K = 2" (Figure 3-5). Press EQN, ENTER to return to display in Figure 3-3.
5. Key VARS for definition of variables (Figure 3-6), then NXT to see units (Figure 3-7; next page). Press the white key "A" to select and display SI units, or "B" to select English units. NXT returns display in Figure 3-6. Key EQN to show display in Figure 3-4.

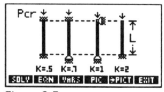
Figure 3-5

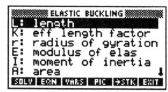
Figure 3-6

6. Key SOLV to begin solution (Figure 3-8)

Figure 3-7

Figure 3-8

7. Enter data by first keying number (Table 3-1), then pressing the white key under the variable displayed at the bottom of the view window. For example, key "10", then press the white key "A" to enter data about the beam's length (L). The small window for the variable darkens when data has been entered. Next, key "2", then press the white key "B" to enter data about beam "Eff. Length Factor". Key remainder of data in a similar way. (Hint: Data for "E" is keyed as: "2.1", EEX, "8", then press the white key "D").
8. To solve the equation, press NXT, then LS, PCR (corresponding white key is "A"). In a moment or two, the solution:

"Pcr = 101.7 kN"

is displayed at the bottom of the view window (Hint: Precision for answer display depends on how the HP48G is configured - see Chapter 1).
9. Press VAR to leave the display for equation solving and display the LOCAL VARIABLES in the "EXER" subdirectory. LS, CLEAR clears the view window.
10. To remove LOCAL VARIABLES, key: LS, { }, then press each white key corresponding to each LOCAL VARIABLE. Key NXT to display additional LOCAL VARIABLES on "page" 2. When all have been listed in the view window, ENTER, LS, PURG erases them.

Exercise 2: Beam Deflection

Problem Statement

Using data in Table 3-2, what is the maximum deflection (cm) of a steel beam 4 cm x 10 cm subjected to the loads P and w?

Table 3-2: Beam Data				
Beam	Factor	Symbol	SI units	Eng. units
6	length	L	m	ft
2.1×10^8	modul. of elasticity	E	kPa	psi
5.33×10^5	moment of inertia	I	mm^4	in^4
3	dist. to point load	a	m	ft
4	point load	P	kN	lbf
0.8	distributed load	w	kN/m	lbf/ft
3	dist. along beam	x	m	ft
notes: $I = bh^3/12$				

Solution (SI units)

(Hint: Screen displays analogous to those in Exercise 1 are not shown.)

Key the following steps:

1. RS, EQ LIB, ENTER, ▼, ▼ (to highlight "Simple Deflection"), ENTER (see equation), PIC (to see model; Figure 3-9), VARS (to see variables), NXT (to select units), NXT (to return to variable list), EQN (Hint: be patient), SOLV (Hint: be patient).

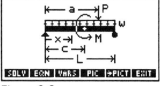

Figure 3-9

2. Enter data for "L", "E" and "I", NXT, enter data for "A", "P", "w", NXT, enter data for "x".

3. LS, Y (Hint: be patient), read:

"y = -28.14 cm"

4. Purge LOCAL VARIABLES for next exercise.

Exercise 3: Electrical Resistance

Problem Statement

What is the equivalent resistance (Rp; Ω) of a network of three resistors in parallel with values of 5 Ω, 10 Ω and 20 Ω?

Solution

(Hint: This solution is a two step process because the HP48G model evaluates only two resistors at a time.)

Key the following steps:

1. RS, EQ LIB, ▼ (to highlight "Electricity"), ENTER, ▼ (5 times, to highlight "Series and Parallel R"), ENTER, PIC (to show models; Figure 3-10; the one on the right is appropriate for this exercise), EQN, NEQN (to show equation for parallel resistors), SOLV.

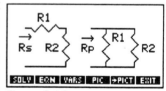
Figure 3-10

(Hint: Quoted statements like, "R1", "R2", etc., refer to variables displayed at the bottom of the view window.)

2. For first solution, key: 5, "R1", 10, "R2", LS, "RP", read:

"Rp: 3.33 Ω"

3. For next solution, key: 3.33, "R1", 20, "R2", LS, "RP", read:

"Rp: 2.85 Ω"

4. Purge LOCAL VARIABLES for next exercise.

Exercise 4: Capacitive Energy

Problem Statement

What is the energy (E; Joules) stored in a 10^6 μF capacitor with an applied voltage of 20 V?

Solution

Key the following steps:

1. RS, EQ LIB, ▼ (to highlight "Electricity"), ENTER, ▼ (8 times to highlight "Capacitive Energy"), ENTER, SOLV, 10, EEX, 6, "C", 20, "V", LS, "E" , read:

2. Purge LOCAL VARIABLES for next exercise.

Exercise 5: Fluid Dynamics

Problem Statement

What is the average velocity (vavg; m/s) and the Reynolds Number (Re; dimensionless) for water flowing in a full pipe under the conditions listed in Table 3-3 (next page)?

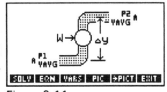
Figure 3-11

Chapter 3 - Solved Problems

Table 3-3: Flow Data

Data	Factor	Symbol	SI units	Eng. units
8	pipe diameter	D	cm	in
0.02	volume flow rate	Q	m^3/s	ft^3/min
1000	density	ρ	kg/m^3	lb/ft^3
0.003	dynamic viscosity	μ	Pa•s	$lbf•s/ft^2$

Solution (SI units)

Key the following steps:

1. RS, EQ LIB, ▼, ▼ (to highlight "Fluids"), ENTER, ▼ (3 times to highlight "Flow in Full Pipes"), ENTER, PIC (to see model, Figure 3-11), EQN (Hint: be patient), SOLV, NXT, 8, "D", NXT, 10, EEX, 3, "ρ", .02, "Q", 10, EEX, 3, +/-, "μ", NXT, LS, "RE" (see interim solutions, then read, "Re: 3.18E5"), NXT, NXT, NXT, LS, "VAV". See interim solutions, then read:

"vavg: 3.98 m/s"

4. Purge LOCAL VARIABLES for next exercise.

Exercise 6: Fluid Pressure

Problem Statement

What is the maximum pressure change (ΔP; kPa) across a 10 HP pump under the conditions listed in Table 3-4 (next page)?

Chapter 3 - Solved Problems

Table 3-4				
Data	Factor	Symbol	SI units	Eng. units
0	height change	ΔY	m	ft^2
100	initial area	A1	cm^2	in^2
100	final area	A2	cm^2	in^2
1000	density	ρ	kg/m^3	lb/ft^3
20	mass flow rate	M	kg/s	lb/min
0	head loss	HL	m^2/s^2	ft^2/s^2
7460	power input	W	W	hp

Solution (SI units)

Key the following steps:

    1. RS, EQ LIB, ▼, ▼ (to highlight "Fluids"), ENTER, ▼, ▼ (to highlight "Flow with Losses"), ENTER, PIC (to see model (Figure 3-11)), SOLV, 0, "ΔY", NXT, 100, "A1", 100, "A2", 1000, "ρ", 20, "M", 0, "HL", 7460, "W", NXT, NXT, LS, "ΔP". See interim solutions, then read:

$$\text{"}\Delta P: 373 \text{ kPa"}$$

    4. Purge LOCAL VARIABLES for next exercise.

Exercise 7: Wind Force

Problem Statement

    What is the maximum force (F; N) because of wind on a 1.2 m by 2 m window under the conditions listed in Table 3-5 (next page)?

Chapter 3 - Solved Problems

Table 3-5:

Data	Factor	Symbol	SI units	Eng. units
0	initial pressure	P1	kPa	psi
0	height change	ΔY	m	ft
27.78	initial velocity	V1	m/s	ft/s
0	final velocity	V2	m/s	ft/s
1.2	density (of air)	ρ	kg/m²	lb/ft²

Solution (SI units)

Key the following steps:

1. RS, EQ LIB, ▼, ▼ (to highlight "Fluids"), ENTER, ▼ (to highlight "Bernoulli Equation"), ENTER, PIC (to see model; Figure 3-12), SOLV, 0, "P1", 0, "ΔY", NXT, 27.78, "V1", 0, "V2", NXT, 1.2, "ρ", NXT, NXT, LS, "ΔP", read:

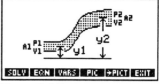

Figure 3-12

"ΔP: 0.463 kPa"

2. Force on window = (ΔP)(A) = (463)(1.2)(2) = 1111 N.
3. Purge LOCAL VARIABLES for next exercise.

Exercise 8: Drag Force

Problem Statement

What is the largest diameter sphere that can move under water and have no greater drag force than 400 N, with the conditions listed in Table 3-6 (next page)?

Table 3-6:

Data	Factor	Symbol	SI units	Eng. units
0.4	drag coefficient	Cd	(none)	(none
1000	fluid (water) density	ρ	Kg/m^3	lb/ft^3
10	velocity	v	m/s	ft/s
400	drag force	F	N	lbf

Solution (SI units)

Key the following steps:

1. RS, EQ LIB, ▼ (3 times to highlight "Forces and Energy"), ENTER, ▼ (5 times to highlight "Drag Force"), ENTER, SOLV, .4, "CD", 1000, "ρ", 10, "V", 400, "F", LS, "A", read:

"A: 200 cm^2"

2. Diameter (D; cm) of sphere = $\sqrt{A \cdot 4/\pi}$ = 15.9
3. Purge LOCAL VARIABLES for next exercise.

Exercise 9: Isothermal Gas Expansion

Problem Statement

What work (W; Joules) is required to expand isothermally 5 kg of air at 23 °C from a pressure of 100 kPa to one of 800 kPa, under the conditions described in Table 3-7 (next page)?

Table 3-7:

Data	Factor	Symbol	SI units	Eng. units
100	initial volume	Vi	L	ft³
800	final volume	Vf	L	ft³
23	temperature	T	°C	°F
172.59	number of moles	N	gmol	lbmol
5	mass	M	kg	lb
28.97	molecular weight	MW	g/gmol	lb/lbmol

notes: 1. N = M/MW
2. For an isothermal condition, Vf/Vi = Pi/Pf.

Solution (SI units)

Key the following steps:

1. RS, EQ LIB, ▼ (4 times to highlight "Gases"), ENTER, ▼, ▼ (to highlight "Isothermal Expansion"), ENTER, SOLV, 100, "Vi", 800, "Vf", 23, "T", 172.59, "N", LS, "W", read:

"W: 8.837E5 J"

2. Purge LOCAL VARIABLES for next exercise.

Exercise 10: Heat Transfer

Problem Statement

What is the steady state heat transfer rate through a window made of two panes of glass separated by an air space, as defined by data in Table 3-8 (next page)? (Hint: Ignore thermal insulation of the glass itself)

Chapter 3 - Solved Problems

Table 3-8				
Data	Factor	Symbol	SI units	Eng. units
20	hot temperature	TH	°C	°F
-10	cold temperature	TC	°C	°F
4	area	A	m^2	ft^2
10	convect. coef. 1	H1	$W/m^2 \cdot K$	$BTU/h \cdot ft^2 \cdot °F$
10	convect. coef.. 3	H3	$W/m^2 \cdot K$	$BTU/h \cdot ft^2 \cdot °F$
0	length 1	L1	cm	in
1	length 2	L2	cm	in
0	length 3	L3	cm	in
1	conductivity 1	K1	$W/m \cdot K$	$BTU/h \cdot ft \cdot °F$
2	conductivity 2	K2	$W/m \cdot K$	$BTU/h \cdot ft \cdot °F$
1	conductivity 3	K3	$W/m \cdot K$	$BTU/h \cdot ft \cdot °F$
notes: K1 and K3 must be greater than zero.				

Solution (SI units)

Key the following steps:

1. RS, EQ LIB, ▼ (5 times to highlight "Heat Transfer"), ENTER, ▼ (4 times to highlight "Conduction and Convection"), ENTER, PIC (to see model; Figure 3-13), SOLV, 20, "TH", 10, +/-, "TC", 4, "A", 10, "H1", 10, "H3", NXT, 0, "L1", 1, "L2", 0, "L3", 1, "K1", 2, "K2", 1, "K3", NXT, LS, "Q", read:

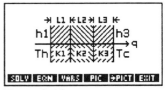

Figure 3-13

"q: 585.4 W"

2. Purge LOCAL VARIABLES for next exercise.

Exercise 11: Projectile Motion

Problem Statement

What is the maximum height (Ymax; m) and the height of a projectile (Yx; m) 300 m from its launch site under the conditions defined in Table 3-9?

Table 3-9				
Data	Factor	Symbol	SI units	Eng. units
0	initial X-position	Xo	m	ft
0	initial Y-position	Yo	m	ft
60	initial angle	θ0	degrees	degrees
50	initial velocity	Vo	m/s	ft/s

Solution (SI units)

(Hint: Range (R; m) is calculated first. Maximum height is then calculated for a distance of R/2 from origin.)

Key the following steps:

1. EQ LIB, ▼ (7 times to highlight "Motion"), ENTER, ▼, ▼ (to highlight "Projectile Motion"), ENTER, PIC (to see model; Figure 3-14), SOLV, 0, "xo", 0, "yo", 60, "θ0", 50, "Vo", LS, "R", read:

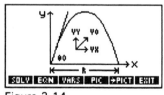

Figure 3-14

"R: 220.8 m"

(Hint: Maximum height occurs at a distance of: 220.8/2 = 110.4 m)

2. To calculate maximum height: NXT, NXT, 110.4, "x", LS, "y". See interim solutions, read:

"y: 95.6 m"

3. To calculate height at 300 m: 300, "x", LS, "y". See interim solutions, read:

"y: -186.6 m"

4. Purge LOCAL VARIABLES for next exercise.

Exercise 12: Linear Motion

Problem Statement

A vehicle traveling at 100 km/h (Vo) brakes and decelerates at 4 m/s^2. How much farther does it travel before it stops?

Solution

(Hint: 100 km/h = 27.78 m/s)

Key the following steps:

1. EQ LIB, ▼ (7 times to highlight "Motion"), ENTER, ("Linear Motion" is highlighted), ENTER, SOLV, (for initial position) 0, "xo", (for initial velocity) 27.78, "Vo", (for deceleration) -4, "A", (for final velocity) 0, "V", (for final position) LS, "X", read:

"x: 96.5 m"

4. Purge LOCAL VARIABLES for next exercise.

Exercise 13: Terminal Velocity

Problem Statement

What velocity does a ball reach as it falls through water under the conditions defined in Table 3-10?

	Table 3-10			
Data	Factor	Symbol	SI units	Eng. units
0.4	drag coefficient	Cd	(none)	(none)
1000	fluid density	ρ	kg/m^3	lb/ft^3
706.9	projected area	A	cm^2	in^2
10	mass	m	kg	lb
notes: $A = \pi D^2/4$				

Solution (SI units)

Key the following steps:

1. EQ LIB, ▼ (7 times to highlight "Motion"), ENTER, ▼ (5 times to highlight "Terminal Velocity"), ENTER, SOLV, .4, "CD", 1000, "ρ", 706.9, "A", 10, "M", LS, "V", read:

"v: 2.63 m/s"

4. Purge LOCAL VARIABLES for next exercise.

Exercise 14: Mass-Spring Oscillation

Problem Statement

What is the oscillation frequency (f; Hz) period (T; s) and angular frequency (ω; r/s) of a 10 kg mass suspended on a spring with a "spring constant" (K) of 600 N/m?

Solution

Key the following steps:

1. EQ LIB, ▲ (6 times to highlight "Oscillations"), ENTER, ""Mass-Spring System" highlighted), ENTER, SOLV, 600, "K", 10, "M", LS, "T", read:

$$\text{"T: 0.81 s"}$$

2. LS, "ω", read:

$$\text{"ω: 7.75 r/s"}$$

3. LS, "F", read:

$$\text{"f: 1.23 Hz"}$$

4. Purge LOCAL VARIABLES for next exercise.

Exercise 15: Plane Geometry

Problem Statement

What is the area (A; cm^2), circumference (C; cm), x, y-moment of inertia (I; mm^4) and polar moment of inertia (J; mm^4) of a 10 cm diameter circle?

Solution

Key the following steps:

1. EQ LIB, ▲ (5 times to highlight "Plane Geometry"), ENTER, ("Circle" highlighted), ENTER, PIC (to see model; Figure 3-15), SOLV, 5, "R", LS, "A", read:

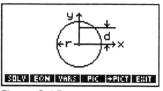
Figure 3-15

"A: 78.54 cm^2"

2. LS, "C", read:

"C: 31.42 cm"

3. LS, "I", read:

"I: 4.909 x 10^6 mm^4"

4. LS, "J", read:

"J: 9.818 x 10^6 mm^4"

4. Purge LOCAL VARIABLES for next exercise.

Exercise 16: Solid Geometry

Problem Statement

What is the volume (V; cm^3) and surface area (A; cm^2) of a 20 cm diameter sphere?

Chapter 3 - Solved Problems   57

2. Purge LOCAL VARIABLES for next exercise.

Exercise 19: Shear Stress (Mohr's Circle)

Problem Statement

What is the maximum principle stress (σ_1; kPa), the maximum shear stress (τ; kPa) and the angle of maximum normal stress (p_1; degrees) when the normal x-stress (σ_x) is 6000 kPa, the normal y-stress (σ_y) is -4000 kPa and the shear stress (τ_{xy}) is 8000 kPa?

Solution

Key the following steps:

1. EQ LIB, ▲ (2 times to highlight "Shear Analysis"), ▼ (3 times to highlight "Mohr's Circle"), ENTER, PIC (to show model; Figure 3-19), 6000, "σ_x", 4000, +/-, "σ_y", 8000, "τ_{xy}", NXT, LS, "σ_1", read:

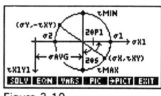

Figure 3-19

"σ_1: 10.433 x 10^3 kPa"

2. LS, "θ_{p1}", read:

"θ_{p1}: 29.00 °"

3. NXT, LS, "τ_{MA}", read:

"τ_{max}: 9.433.98 kPa"

4. Purge LOCAL VARIABLES for next exercise.

Exercise 20: Roots of a Polynomial

(Hint: Configure display to "FIX 4")

Problem Statement No. 1

Find the roots of:
$$X^2-X-6=0$$

Solution

Key the following steps:

1. RS, SOLVE, ▼, ▼ (to highlight "Solve poly..."), ENTER, LS, [ ], 1, SPC, 1, +\-, SPC, 6, +\-, OK, SOLV, read:

$$\text{"[-2 3]"}$$

indicating the roots are: $x = -2$ and $y = 3$.

Problem Statement No. 2

Find the one complex root and the complex conjugate of:
$$X^2-X+6=0$$

Solution

Key the following steps:

1. RS, SOLVE, ▼, ▼ (to highlight "Solve poly..."), ENTER, LS, [ ], 1, SPC, 1, +\-, SPC, 6, OK, SOLV, read:

$$\text{"(.5, 2.3979)"}$$

indicating one complex root is: $X = .5 + 2.3979i$. The complex conjugate must then be: $X = .5 - 2.3979i$.

Chapter 3 - Solved Problems   59

Problem Statement No. 3

Find the roots of:

$$X^3 - X - 6 = 0$$

Solution

Key the following steps:

1. RS, SOLVE, ▼, ▼ (to highlight "Solve poly..."), ENTER, LS, [ ], 1, SPC, 0, SPC, 1, +\-, SPC, 6, +\-, OK, SOLV, read:

$$\text{``}(-1, 1.4142)\text{''}$$

2. EDIT, read:

$$X1 = -1 + 1.4142$$

3. RC, read:

$$X2 = -1 - 1.4142$$

4. RC, read:

$$X3 = 2$$

Exercise 21: Solving Equations

Problem Statement

Solve the equations:

$$\tan x = 0$$

and

$$x^2 + e^x = 0$$

Solution

Key the following steps:

1. RS, SOLVE, ("Solve equation" is highlighted), ENTER, TAN, α, X, RC, LS, =, 2, OK, SOLVE, read:

"x: 63.4349"

2. ▼, α, X, y^x, +, LS, e^x, α, X, OK, SOLVE, read:

"x: -.35173"

Exercise 22: First-order Differential Equation

Problem Statement

Find y(2) for:

$$y' = y\ SIN\ t$$

if

$$y(0) = 1$$

Solution

Key the following steps:

1. RS, SOLVE, ▼ (to highlight "Solve diff eq..."), ENTER, LS, EQUATION, α, Y, SIN, α, T, RC, ENTER, ▼, α, T, OK, RC, 2, OK, RC, 1, OK (see Figure 3-20) SOLVE, EDIT, read:

Figure 3-20

"1.0355"

4. CANCEL, CANCEL

Exercise 23: Solve a System of Linear Equations

Problem Statement

Solve the system:
$$2x+y+z=6$$
$$x-y-z=0$$
$$y+2z=0$$

Solution

Key the following steps:

1. RS, SOLVE, ▼ (3 times to highlight "Solve lin sys..."), ENTER, EDIT, 2, ENTER, 1, ENTER, 1, ENTER, ▼, 1, ENTER, 1, +/-, ENTER, 1, +/-, ENTER, 0, ENTER, 1, ENTER, 2, ENTER, ENTER, ▼, EDIT, 6, ENTER, 0, ENTER, 0, ENTER, ENTER, ▼, SOLVE, read:

"2 4 -2"

indicating that: $x=2$, $y=4$ and $z=-2$.

Exercise 24: Second-order Differential Equation

Problem Statement

Solve:
$$\ddot{y}+2\dot{y}+10y=2t$$

for y(2) if y(0) = 0 and $\dot{y}=(0)=10$

Let $\dot{y}=x$, $\dot{x}+2x+10y=2t$

or $\dot{x}=-2x-10y+2t$
$\dot{y}=x$

Chapter 3 - Solved Problems

Solution

Key the following steps:

1. RS, MATRIX, 2, +/-, ENTER, 10, +/-, ENTER, ▼, 1, ENTER, 0, ENTER, ENTER, ', α, A, STO, LS, [ ], 2, SPC, 0, ENTER, ', α, B, STO

2. RS, SOLVE, ▼ (to highlight "Solve diff eq..."), ENTER, α, A, *, α, W, +, α, B, *, α, T, ENTER, α, T, OK, RC, 2, OK, α, W, OK, LS, [ ], 10, SPC, 0, OK, RC, .001, OK, ▲, RC (see Figure 3-21), SOLVE (Hint: be patient), EDIT, read:

Figure 3-21

"1.60216
.24153"

to indicate that

$$x(2) = \dot{y}(2) = 1.60216$$

and

$$y(2) = 0.24153$$

Chapter 4

Special Applications

- Matrix Operations: Constructing, Storing, Recalling and Editing

- Examples of Matrix Operations

- Solving a System of Linear Equations

- Applications Using Complex Numbers and Vectors

- Simultaneous Solutions of Non-linear Equations

- Sample Problems

Notes

Chapter 4 - Special Applications

This chapter presents three kinds of special applications. The first describes techniques for performing matrix operations, then shows how to solve a system of linear equations. The second illustrates the use of complex numbers and vectors. The third shows how to use the HP-48G for finding simultaneous solutions of non-linear equations. These operations are often complicated and challenging, but the HP-48G makes them easy.

Matrix Operations

As for so many operations with the HP-48G, there are often several, equally good ways to achieve the same computational end. Which one is "best", depends on the user's experience, habits and intuition. When performed correctly, of course, all lead to the same results.

The chapter "Basic Statistics" introduced techniques for constructing and manipulating an array of data either as a vector (a single column of data) or as a matrix (data arranged in rows and columns). Sample problems in this section use many of the same procedures.

Matrix operations in this chapter are demonstrated within a subdirectory called, "MAT". To construct this subdirectory, press: RS, MEMORY, NEW, ▼, α, α, M, A, T, α, ENTER, ✓CHK, OK, CANCEL. To enter the subdirectory, press: the white key marked "A" ("MAT").

Problem Statement

Construct and store the following matrix in the X STACK REGISTER

Chapter 4 - Special Applications

(hint: display mode is set at "Std" 0 (zero)):

$$\begin{bmatrix} 3 & 2 \\ 3 & 4 \end{bmatrix}$$

Solution

Step   Press

1   RS, MATRIX, 3, SPC, 2, ENTER, ▼, 3, SPC, 4, ENTER
     (Figure 4-1), ENTER (Figure 4-2)

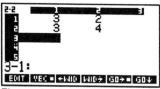
Figure 4-1

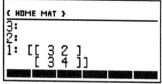
Figure 4-2

Problem Statement

Construct and enter the second matrix in a similar way.

$$\begin{bmatrix} 4 & 3 \\ 1 & 4 \end{bmatrix}$$

Figure 4-3 shows how the two matrix tables appear in the view window. To multiple, add or subtract the two matrices, press the appropriate function key. For example, Figure 4-4 shows the results of multiplication of the matrices shown in Figure 4-3.

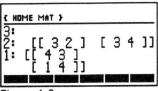
Figure 4-3

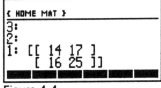
Figure 4-4

Other arithmetic operations using matrices are performed with analogous procedures. For example, calculating the inverse of a matrix requires generating the matrix, then pressing the "1/x" key. To find the determinant of a matrix, generate the matrix, then press: MTH, MATR, NORM, NXT, DET. To find the eigenvalues if a matrix, press: MTH, MATR, NXT, EGVL.

Storing, Recalling and Editing a Matrix

Once a matrix has been generated, as shown in Figure 4-2, it can be stored, then recalled for different operations. Beginning with Figure 4-2, for example, the matrix is stored as "MAT1" by pressing: ', α, α, M, A, T, 1, α, ENTER, STO. To recall the matrix table, press the white key marked "A" to select "MAT1". To edit the matrix entitled, "MAT1":

Step	Press
1	Press white key to select "MAT1", then press ▼.
2	Use cursor control keys to highlight cell whose data are to be edited

3 Key new data, press ENTER

4 Repeat Steps 2 and 3, if necessary, and edit other data cells

5 Press ENTER to exit edit status and return to display window

To remove a matrix from storage, key: ', (select matrix), ENTER, LS, PURG.

Examples of Matrix Operations

Problem Statement

Generate and store the following matrices:

$$A = \begin{bmatrix} -1 & 2 & 2 \\ 2 & 2 & 2 \\ -3 & -6 & -6 \end{bmatrix} \quad B = \begin{bmatrix} 3 & 0 & 1 \\ 0 & 2 & 0 \\ 5 & 0 & -1 \end{bmatrix}$$

Solution

1.

For A: RS, MATRIX, 1, +/-, SPC, 2, SPC, 2, ENTER, ▼, 2, SPC, 2, SPC, 2, ENTER, 3, +/-, SPC, 6, +/-, SPC, 6, +/-, ENTER, ENTER, ', α, α, M, A, T, A, α, STO

For B: RS, MATRIX, 3, SPC, 0, SPC, 1, ENTER, ▼, 0, SPC, 2, SPC, 0, ENTER, 5, SPC, 0, SPC, 1, +/-, ENTER, ENTER, ', α, α, M, A, T, B, α, STO

Chapter 4 - Special Applications

2. To multiply MATA times MATB, press the white key marked "B" (to bring MATA to the view window), then press white key "A" (for MATB), then press the multiplication key to see that:

$$A \cdot B = \begin{bmatrix} -1 & 2 & 2 \\ 2 & 2 & 2 \\ -3 & -6 & -6 \end{bmatrix} \begin{bmatrix} 3 & 0 & 1 \\ 0 & 2 & 0 \\ 5 & 0 & -1 \end{bmatrix} = \begin{bmatrix} 7 & 4 & -3 \\ 16 & 4 & 0 \\ -39 & -12 & 3 \end{bmatrix}$$

3. To multiply MATB times MATA, first recall MATB, then MATA, then press the multiplication key to see that:

$$B \cdot A = \begin{bmatrix} 3 & 0 & 1 \\ 0 & 2 & 0 \\ 5 & 0 & -1 \end{bmatrix} \begin{bmatrix} -1 & 2 & 2 \\ 2 & 2 & 2 \\ -3 & -6 & -6 \end{bmatrix} = \begin{bmatrix} -6 & 0 & 0 \\ 4 & 4 & 4 \\ -2 & 16 & 16 \end{bmatrix}$$

4. To find the determinant of MATB, first recall MATB, then press MTH, MATR, NORM, NXT, DET to see that:

$$\begin{vmatrix} 3 & 0 & 1 \\ 0 & 2 & 0 \\ 5 & 0 & -1 \end{vmatrix} = -16$$

(Hint: Press VAR to return to display of local variables in HOME MAT)

5. To determine the eigenvalues for MATA, first recall MATA, then press MTH, MATR, NXT, EGVL to read: "[-3, -2, 0]". (Note: The zero may appear as "-2 x 10^{-14}" or "-1.86...x 10^{-14}" or some similarly small number, depending on select display format and rounding errors.)

6. To calculate the reciprocal of MATB, first recall MATB, the press "1/x" to see that:

$$B^{-1} = \begin{bmatrix} .125 & 0 & .125 \\ 0 & .5 & 0 \\ .625 & 0 & -.375 \end{bmatrix}$$

(Hint: Be sure the display format is set to show an adequate number of digits to the right of the decimal for accuracy.) Note: The reciprocal of MATA doesn't exist because its determinant is zero.

Solving a System of Linear Equations

Problem Statement

Evaluate the unknowns for:

(hint: display mode is set at "FIX" 2)

Chapter 4 - Special Applications

$$a + 2.1b - 3.4c = 9.8$$
$$3.9a - 1.9b + 0.9c = -4.1$$
$$1.9a + 3.2b - c = 14.8$$

Solution

Step	Press	
1	(construct a matrix for the vector on the right-hand side of the equations) RS, MATRIX, 9.8, SPC, 4.1, +/-, SPC, 14.8, ENTER, ENTER (Figure 4-5)	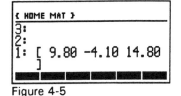 Figure 4-5

2   (construct a matrix for equation coefficients) RS, MATRIX, 1, SPC, 2.1, SPC, 3.4, +/-, ENTER, ▼, 3.9, SPC, 1.9, +/-, SPC, .9, ENTER, 1.9, SPC, 3.2, SPC, 1, +/-, ENTER, ENTER (Figure 4-6), ÷ (Figure 4-7)

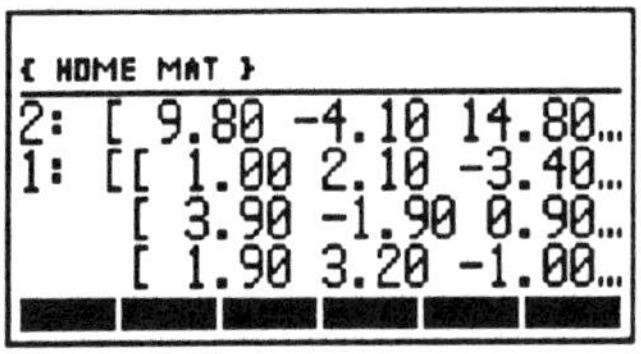

Figure 4-6

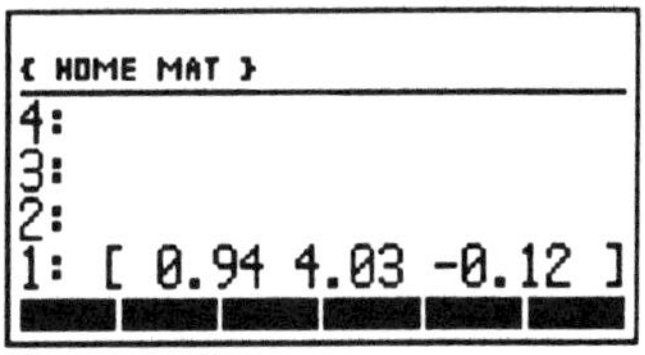

Figure 4-7

The calculation shows that:

	Display Format	
	FIX, 2	STD
a	0.94	0.9393
b	4.03	4.031
c	-0.12	-0.1164

Applications Using Complex Numbers and Vectors

Applications often require the use of complex numbers. When solving a quadratic equation, for example, the roots may be real or complex, depending on its coefficients. Engineering applications may also require complex numbers or vector expressions.

To input the complex number, 2+3i, press: LS, ( ), 2, SPC, 3, ENTER. To input a second complex number, 1-4i, press: LS, ( ), 1, SPC, 4, +/-, ENTER. These entries can be multiplied or divided by pressing the appropriate function key.

If "Rectangular" has been selected as the "COORD SYSTEM" in "CALCULATOR MODES", the complex number, 2+3i, will be displayed as "(2,3)" if the "NUMBER FORMAT" is "Std". It will appear as, "(2.00,3.00)" if the "NUMBER FORMAT" is selected to be "Fix 2". If the "COORD SYSTEM" is chosen to be "Polar", 2+3i will be displayed as, "(3.61,∆56.31)", the magnitude and angle.

To input the complex number, 3∠45, press: LS, ( ), 3, SPC, RS, ∆, 45, ENTER. If the "NUMBER FORMAT" is set at "Std" and the "COORD SYSTEM" is set at "Polar", the number will be displayed as,

"(3,∡45)". If the "NUMBER FORMAT" is set at "Fix 2" and the "COORD SYSTEM" is set at "Rectangular", the number will appear as, "(2.12,2.12)".

To input a vector 2i + 3j, press: LS, [ ], 2, SPC, 3, ENTER. With "NUMBER FORMAT" set in "CALCULATOR MODES" at "Fix 2" and "COORD SYSTEM" selected to be "Rectangular", the entry is displayed as, "[2.00 3.00]". It appears as, "[3.61 ∡56.31]" when "COORD SYSTEM" is "Polar".

When a second vector, 2i-4j + 3k, is entered, the cross or dot products are found by pressing MTH, VECTR, then pressing the white key corresponding to the appropriate function, or determining the absolute value by selecting "ABS".

Simultaneous Solutions for Non-linear Equations - An Example

Background

The Australian brush turkey (*Alectura lathami*) uses a unique technique for incubating its eggs. Instead of brooding them, as do many other birds, these animals deposit their eggs in a cleverly crafted mound of fermenting vegetation which they scratch together from leaves, twigs and other debris on the forest floor.

Such mounds are typically about a meter high and about five meters in diameter. These birds meticulously groom the pile of rubble during the period of egg incubation so that oxygen, carbon dioxide and water vapor partial pressures in the mound remain in safe limits for the developing embryos. They also fine-tune the geometry of the mound so that its rate of heat production from natural fermentation precisely balances that of its heat loss to the surrounding environment (Figure 4-8, next page). The system is tailored such that the incubating eggs are held close to an optimum temperature of about 33C.

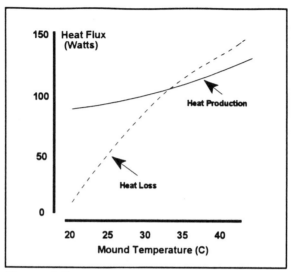

Figure 4-8

There are passive processes at work in the mound that help stabilize its temperature. When nest temperature rises above about 33C, then, as shown in Figure 4-8, heat loss from the mound transiently exceeds that of heat production, and the nest cools. The reverse is true were the nest to fall below about 33C. Within the limits of incubation requirements, these birds have constructed a self-regulating temperature environment for their offspring.

The energetics of the mound enclosure are even more remarkable. Many of its net thermodynamic effects approximate those of living animals. A nest temperature of about 33C, for example, corresponds closely to a normal average skin surface temperature for humans and other warm blooded animals when they are exposed to thermoneutral environments.

Also, the steady state heat loss rate of about 110 Watts for the mound at a temperature of about 33C is just a little less than that from the human body surface (about 120 Watts) when someone is sitting quietly and resting. In a steady state, the net effects of heat production by fermentation and heat transfer by thermal conduction in the mound, along with evaporation and convection at its surface must approximate those of heat transfer and loss in the human body, even though basic control mechanisms are quite different.

The coordinates at which the functions of heat production and heat loss cross in Fig. 1 mark the value of optimum nest temperature and the associated steady state heat flux rate from the mound. Heat production rates ($H_{production}$; Watts) as a function of mound temperature (T; deg. Celsius) are approximated by:

$$H_{production} = 55.64 \, e^{0.021T} \tag{1}$$

Heat loss rates (H_{loss}; Watts) as a function of mound temperature (T; deg. Celsius) are approximated by:

$$H_{loss} = 208.1 \, \ln(T) - 617.4 \tag{2}$$

When the nest is thermally stable,

$$H_{loss} = H_{production}$$

Chapter 4 - Special Applications

so that equations (1) and (2) give

$$208.1 \ln(T) - 617.4 = 55.64\, e^{0.021T} \qquad (3)$$

The graphical solution (Figure 4-8) of equation (3) for these simultaneous processes for either nest temperature or steady state heat production is much simpler than is the algebraic one. In fact, the relationship cannot be solved algebraically. But, it can be solved by iteration, which is a simple job with the HP-48G.

(HINT: the following solution is made in a subdirectory called, "NEST" with display mode set to display one digit to the right of the decimal ("FIX 1"))

(Reminder: "LS" = left shift; "RS" = right shift; "RC" = right cursor)

Step   Press

1   LS, EQUATION, 208.1, RS, LN, α, T, RC, -, 617.4, LS, =, 55.64, LS, e^x, .021, α, T, RC, ENTER

2   (store the equation) ', α, H, STO

3   RS, SOLVE, "Solve equation..." (highlighted), OK , CHOOS, "H:'208.1*LN(T..." (highlighted), OK, SOLVE (read "T:33.3")

4   CANCEL (to exit SOLVE operation)

Chapter 4 - Special Applications

The solution to the non-linear equation (3) is: $T = 33.3C$. The heat loss rate is then found from equation (2) to be 112.1W.

Rather than solving for T in equation (3), an expression could have been found for T from equation (1) and a second expression from equation (2). Equating the T's and letting

$$H_{production} = H_{loss} = H,$$

provides a non-linear equation for H that could be solved with a procedure similar to that used to find T. An alternative is to let the HP48 provide a solution for equation (2) by:

Step   Press

1   LS, EQUATION, α, H, LS, =, 208.1, RS, LN, α, T, RC, -, 617.4, ENTER

2   (store equation) ', α, E, 1, STO

3   RS, SYMBOLIC, (highlight "Isolate var..."), OK, CHOOS, (highlight "E1"), OK, ▼, α, T, ENTER, OK (Figure 4-9), ▼ (Figure 4-10)

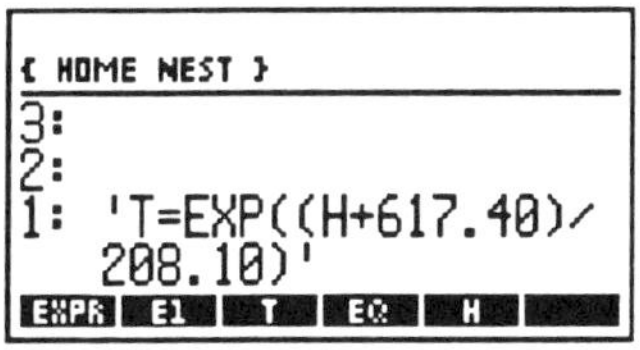

```
{ HOME NEST }
3:
2:
1: 'T=EXP((H+617.40)/
    208.10)'
[EXPR][E1][T][EQ][H]
```
Figure 4-9

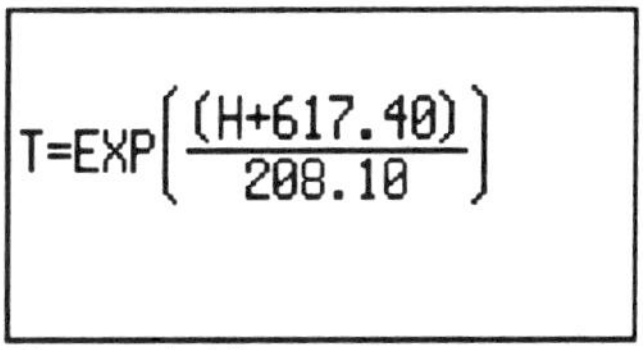

$$T = EXP\left(\frac{(H+617.40)}{208.10}\right)$$
Figure 4-10

Figure 4-10 shows that equation (2) can be written in the form

$$T = e^{\left[\frac{(H+617.4)}{208.1}\right]} \quad (4)$$

Similarly, isolating "T" in equation (1) results in

$$T = \frac{\ln\left(\frac{H}{55.64}\right)}{0.021} \quad (5)$$

To find the steady state heat flux, solve for H in

$$e^{\left[\frac{(H+617.4)}{208.1}\right]} = \frac{\ln\left(\frac{H}{55.64}\right)}{0.021} \quad (6)$$

Following the procedure used to find T from equation (3),

$$H = 111.9 \text{ W}$$

Using this value in equations (4) or (5) gives

$$T = 33.3 \; C$$

The techniques and procedures used to solve the simultaneous non-linear equations (1) and (2) can be applied to many situations in the sciences.

Notes

Chapter 5

Writing Programs

- Basics of Program Structure
- Using Special Characters
- Program Design
- Symbolic Solutions
- Drawing a Flow Diagram
- Constructing and Using ALPHA Characters
- Keying Programs
- Sample Programs

Notes

Chapter 5 - Writing Programs

Basics of Program Structure

The function of a program for the HP-48G is identical to that for a tabletop computer. It provides of series of instructions that are automatically and sequentially completed to arrive either at a numerical or a logical endpoint. Also, the basic structure of the program is the same for both kinds of machines.

Every program has an initial stage of INPUT, a secondary stage of PROCESSING and a final stage of OUTPUT. The INPUT stage allows the user to provide required data either in numerical, or alphabetical form. The PROCESSING stage uses this information to complete specific logic sequences, conditional tests and/or solutions of equations. The OUTPUT stage structures answers with appropriate alphanumeric symbols and units, determines where and how long this information is displayed, and then defines what the program does next. It may start again, end, initialize some other program, or branch in some other way. Even though a program may have many subdirectories, branches and loops, each starts at a specific beginning (input), each performs a specifically designated chain of operations (processing) and each provides answers in a constructed format (output).

There are two steps in writing an HP-48G program. The first is to construct appropriate statements to control INPUT, PROCESSING and OUTPUT using functions accessed from the keyboard and from the calculator's many features in ROM. The second step is to store this code as a suitably named variable in an appropriately named subdirectory. Activating the variable, runs the program. The whole procedure is much easier to do, than it is to describe or read about.

Using Special Characters

Writing text in a program to request data and to display answers is conveniently and easily done with the HP-48G, thanks to its large

84   Chapter 5 - Writing Programs

library of special characters (Figures 5-1 to 5-4) that supplements its ability to generate a complete set of alphabetic and number symbols. The special character menus are accessed by keying RS, CHARS.

Figure 5-1

Figure 5-2

Figure 5-3

Figure 5-4

The options at the bottom of each view window of "-64" and "+64" allow moving among the displays. The option "ECHO" copies the highlighted character to Line 1 of the view window. Display of the special character menu is terminated by pressing NXT, then CANCL, or pressing the ON key to activate its CANCEL function. The CANCEL option is used in the following example.

Example for using Special Characters:

Imagine a program has successfully completed a series of calculations. It is now time to display in the view window the phrase, " X̄ Length = 35 μM" (Figure 5-5).

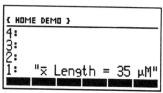
Figure 5-5

Chapter 5 - Writing Programs 85

It is constructed by:

Step	Operation
1	RS, " ", RS, CHARS, RC, ECHO, CANCEL, SPC, α, α, L, LS, α, E, N, G, T, H, SPC, LS, =, 3, 5, α, SPC, RS, CHARS, (use ▼ and RC to highlight "μ"), ECHO, CANCEL, α, M, ENTER.
2	LS, CLEAR

Program Design

There are many ways to design a program. None is right or wrong out of the context of what the program is intended to do. A good case can be made that any program is a good one if it gets the job done accurately, clearly and obligates no more memory in the calculator than necessary. This leaves the user with great discretion in how programs are organized and executed.

This chapter focuses on developing a program for the HP-48G that makes the solution of just one problem - calculating the thermal conductivity of a wall. The associated equation is easily solved either by using the one stored in the EQUATION LIBRARY, or by using the SOLVE function. The purpose of the demonstration in this chapter, though, is to illustrate the processes for designing, constructing and using a program, not just to provide a numerical answer.

Development of the program goes through several stages. The first involves rearranging elements in a simple equation. The second describes the design of a program to solve the equation, including writing a flow diagram for it. Finally, the program is written and a sample problem is used to test it. Although most people would not need a calculator as sophisticated as the HP-48G to do all this, the simplicity of the equation helps focus attention on the processes involved, first in the symbolic algebraic solution, then in writing the program.

The next example shows how to rearrange the equation,

$$q = \frac{(k)(A)}{L}(Th-Tc)$$

to solve for "k". The example then illustrates how to write appropriate INPUT statements, construct the equation for PROCESSING, then display the answer as an OUTPUT with associated phrases and units.

Example for Symbolic Solution

Problem Statement:

Write a program that solves for the thermal conductivity (k; BTU/(h•ft•F) of a wall, given its thickness (ft), surface area (ft^2), steady state heat flux rate (BTU/h) and the temperatures on its warmer inside surface (Th; F) and colder outside surface (Tc; F).

Solution: Step 1

Solve for "k".

The equation,

$$q = \frac{(k)(A)}{L}(Th-Tc)$$

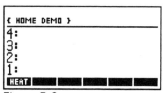
Figure 5-6

has been written and stored as the LOCAL VARIABLE called, "HEAT" in a subdirectory called, DEMO. (Figure 5-6).

Step	Operation
1	The symbolic solution for "k" begins by keying RS SYMBOLIC, then using the ▼ key to highlight "Isolate variable" (Figure 5-7).

Figure 5-7

Figure 5-8

2	Press "OK" to display the "ISOLATE A VARIABLE" menu (Figure 5-8). Next, press "CHOOS" to display the functions stored in the "HOME DEMO" subdirectory, which is only HEAT in this example (Figure 5-9), and which is highlighted by default. Were there other functions, they would be displayed in subsequent lines and highlighted by using the vertical cursor controls.

Figure 5-9

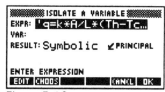
Figure 5-10

3	Press "OK" to display Figure 5-10. Press the ▼ key once, then define the variable ("VAR") by keying: α, LS, K, ENTER (Figure 5-11, next page). The symbol for the variable "k" must be defined as a lowercase letter because it was stated as such when the

expression (displayed currently at the top of menu) was originally written.

Figure 5-11   Figure 5-12

4   Press "OK" to see new expression that defines "k" (Figure 5-12). A new variable, "EXPR", is automatically added to the subdirectory. Store the equation at Line 1 of the view window by keying: ', α, α, K, S, O, L, α, ENTER, STO. (Figure 5-13).

Figure 5-13   Figure 5-14

5. Recall the equation for "k" by pressing "KSOL" (Figure 5-14). See the equation in an EQUATION WRITER format by pressing ▼. (Figure 5-15). Key, CANCEL, CANCEL, LS, CLEAR to return to a blank view window.

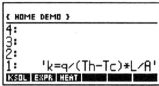

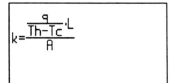

Figure 5-15

Chapter 5 - Writing Programs 89

Solution: Step 2 - Designing, Keying and Using the Program.

A Friendly Warning

Keying a program the first couple of times can be quite a challenge. Keystrokes don't always yield what one hopes for, mysterious symbols seem to come from nowhere, alpha characters get jumbled, and any number of problems develop. With patience and practice, however, it all becomes clear. But, it won't be at first.

There are, however, several safety nets. Follow the keystroke instructions carefully to duplicate the program described in this section. Even if the functions of some program instructions are not understood, the program will still run as long as it is entered correctly into the HP-48G. Also, to get more background information and step-by-step instructions for designing and keying programs, read "Mastering the HP-48G/GX" described elsewhere in this book.

- By all means -
don't abandon efforts to learn programming!

Designing the Program:

Thanks to the simplicity of the equation for calculating "k", the design of a program is also simple, as shown in Figure 5-16 (next page). The program is entitled, "KPRG". When initialized, its INPUT component requests data for heat flow (Q), associated temperatures (Th and Tc), wall thickness (L) and area (A). The program is designed so that the user is reminded of the units appropriate for each data entry. Each keyed number is entered by pressing ENTER. PROCESSING in the program involves calculating "k", and OUTPUT displays the answer with units. All continues automatically, of course, once data are entered.

The program includes some non-essential elements. A tone is generated, for example, to alert the user for each datum input, and

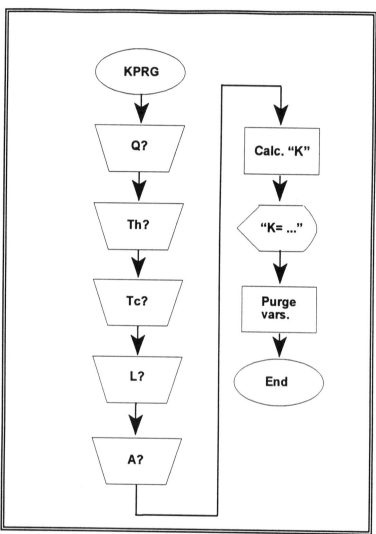

Figure 5-16

tones announce the answer which is displayed for 5 seconds. The program then purges local variables generated by the program, and finally clears the view window. This leaves the user with an uncluttered subdirectory and a blank view window for whatever operation comes next. The program is also designed to use subroutines for repetitive operations, like the sounding of tones and ancillary instructions for storing input data.

Program Listing:

KPRG:
«4 FIX TN
"Heat Flux?"
":Btu/hr:" I 'Q'
STO TN
"Inside Temp?"
":deg F:" I 'Th'
STO TN
"Outside Temp?"
":deg F:" I 'Tc'
STO TN
"Wall Thickness?"
":feet:" I 'L' STO
TN "Wall Area?"":sq. feet:" I 'A'

STO 'Q/(Th-Tc)*L/A'
→NUM 'K' STO TN TN
CLLCD "K = " K + 2
DISP "Btu/(h*ft*F)"
3 DISP 5 WAIT { Q
Th Tc L A } PURGE
CLEAR»

I:
«INPUT OBJ → »

TN:
«2000 .1 BEEP»

Detailed Program Description

The program is written in the subdirectory DEMO. Because of the ease of activating a program and seeing no more "pages" of local variables than necessary, it is an excellent idea to *write each program in an appropriately named subdirectory of its own*. This also allows designations of local variables to be used again in other subdirectories. If there are several programs related to a specific topic, subdirectories can be written within the topic subdirectory.

Chapter 5 - Writing Programs

The instruction "4 FIX" at the beginning of the program sets the number of digits to the right of the decimal for the answer display. "TN" branches the program to the subroutine TN to sound a tone of 2,000 Hz for 0.1 seconds. The phrase, "Heat Flux?", followed by the construction ":Btu/hr:" holds the data request at the top of the view window while the reminder about the correct units is at the bottom of the display. Numerical data are shown to the right of the units designation where they can be edited.

When the number is correct for the data request, it is placed into memory by pressing ENTER. The construction, "I" directs the program to a subroutine where whatever number was entered is received as an input. The construction, " 'Q' STO" stores it in a local variable called "Q". The named variable appears immediately at the bottom of the view window. Other data are requested in a similar way and appropriately defined local variables are constructed for them. This completes the INPUT part of the program.

PROCESSING in the program begins with the instruction, 'Q/(Th-Tc)*L/A' which takes numbers from correspondingly named local variables to complete the calculation of "k". A number is constructed from the equation's operation by " →NUM" and stored in a generated local variable "K" by the phrase " 'K' STO".

OUTPUT in the program begins when two tones are sounded by "TN TN", the display is cleared by "CLLCD", and the alpha phrase "K =" is constructed. The calculated value for "k" is recalled by "K" and added to the previous phrase by "+" and displayed at line 2 of the display (this is not the same location as Line 2 of the STACK). Appropriate units are appended to the display by "Btu/(h*ft*F)" and appear at line 3 of the display. The fully constructed display is held for 5 seconds by "5 WAIT", then local variables generated by the program are purged by "{Q Th Tc L A} PURGE". The calculated value for "k" in the local variable "K" is not purged, but remains at the bottom of the view window so that it can be recalled, if necessary, by the user. The number stored in "K" is updated each time the program is run. In this version of the program, only the most recently calculated "k" is available for review. The last instruction in the program clears the STACK by CLEAR. This short example shows all too few of the many powerful features of the HP-48G for writing programs.

> *Three procedures make all program writing easy:*
> *- practice, practice and practice -*

Keying ALPHA (α) Symbols

Like so many of the HP-48G's operations, there are several ways to achieve the same end. The following suggestions present just one way. With just a little experience, everyone soon finds out what is the easiest for them.

The program for calculating "k" includes ALPHA statements with both upper and lowercase symbols. Table 5-1 gives some guidelines for creating them.

Table 5-1: Generating ALPHA (α) Characters	
Press	**Next Key Produces**
α once	uppercase letter
α twice	continuous string of uppercase letters
α, LS	lowercase letter
α, RS	"Special Character"
α, LS, α, α, α	continuous string of lowercase letters
Hint: After keying α twice, then LS, α toggles between upper and lower case letters until α is pressed again.	

Keying the Program

What's going to happen: Step 1 keys the program instructions. Step 2 stores them as the variable, "KPRG". Step 3 keys the instructions for a subroutine. Step 4 stores them as, "I". Step 5 keys the instructions for the tone. Step 6 stores them as, "TN".

> *Hints:* LS = left shift; RS = right shift; RC = right cursor control; ← = erase key; ' = apostrophe key; ✳ = multiplication key; " " = RS function of subtraction key; : : = RS function of addition key; ( ) = LS function of division key; { } = LS function of addition key; « » = LS function of subtraction key.

Reminder: Write the following program in a subdirectory called, DEMO.

Step   Press
 1   LS, « », 4, LS, MODES, FMT, FIX, α, α, T, N, SPC, T, N, α,
SPC, RS, " ", α , α, H, LS, α, E, A, T, SPC, LS, α, F, LS, α, L, U, X, LS , ←, α, RC,
RS, " ", RS, : :, α, α, LS, α, B, LS, α, T, U, ÷, H, R, LS, α, α, RC, RC, α, I, ', α, Q,
RC, STO, α, α, T, N, α, RS, " ", α , α, I, LS, α, N, S, I, D, E , SPC, LS, α, T, LS, α,
E, M, P, LS , ←, α, RC, RS, " ", RS, : :, α, α, D, E, G, SPC, LS, α, F, α, RC, RC, α,
I, ', α, α, T, LS , α, H, α, , RC, STO, α, α, LS, α, T, N, α, RS, " ", α , α, O, LS,
α, U, T, S, I, D, E, SPC, LS, α, T, LS, α, E, M, P, LS , ←, α, RC, RS, " ", RS, : :, α,
α, D, E, G, SPC, LS, α, F, α, RC, RC, α, I, ', α, α, T, LS , α, C, α, , RC, STO, α,
α, LS, α, T, N, α, RS, " ", α , α, W, LS, α, A, L, L, SPC, LS, α, T, LS, α, H, I, C,
K, N, E, S, S, LS , ←, α, RC, RS, " ", RS, : :, α, α, F, E, E, T, α, RC, RC, α, α, L S,
α, I, α, ', α, L, , RC, STO, α, α, T, N, α, RS, " ", α , α, W, LS, α, A, L, L, SPC,
LS, α, A, LS, α, R, E, A, LS , ←, α, RC, RS, " ", RS, : :, α, α, S, Q, ., SPC, F, E, E,
T, LS, α, α, RC, RC, α, I, ', α, A, RC, STO, α, α, T, N, α, ', α, α, Q, ÷, LS, ( ), T,
LS α, H, -, LS, α, T, LS, α, C, α, RC, ✳, α, α, LS, α, L, ÷, A, α, RC, LS, →NUM,
', α, K, RC, STO, α, α, T, N, SPC, T, N, α, PRG, NXT, OUT, CLLCD, RS, " ", α, α,
K, LS, =, α, RC, α, K, +, SPC, 2, PRG, NXT, OUT, DISP, RS, " ", α, α, B, LS, α,
T, U, ÷, LS, ( ), H, ✳, F, T, ✳, LS, α, F, α, RC, RC, 3, PRG, NXT, OUT, DISP, 5,
PRG, NXT, IN, WAIT, LS, { }, α, α, Q, SPC, T, LS, α, H, SPC, LS, α, T, LS, α, C,
SPC, LS, α, L, SPC, α, A, RC, LS, PURG, LS, CLEAR, ENTER

Chapter 5 - Writing Programs 95

2   ', α, α, K, P, R, G, α, STO
3   LS, « », PRG, NXT, IN, INPUT, PRG, TYPE, OBJ→, ENTER
4   ', α, I, STO
5   LS, « », 2, 0, 0, 0, SPC, ., 1, SPC, PRG, NXT, OUT, NXT, BEEP, ENTER
6   ', α, α, T, N, α, STO

Press VAR to see local variables.

> **In Case of Mistakes**
>
> To review and/or edit program, key: ', KPRG, LS, EDIT, position cursor, make changes, then press ENTER to save program.

Running the Program

Step   *Press*

1   Turn HP-48G on. Press white key under "DEMO" in HOME directory, if necessary, to enter the DEMO subdirectory (Figure 5-17).

2   Press KPRG. Hear two tones and see data request for "Heat Flux" (Figure 5-18). Key: 700, press ENTER

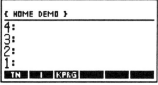
Figure 5-17

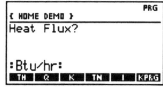
Figure 5-18

96   Chapter 5 - Writing Programs

3   Hear tone and see data request for "Inside Temp" Figure 5-19). Key: 75, press ENTER.

4   Hear tone and see data request for "Outside Temp" (Figure 5-20). Key: 32, press ENTER.

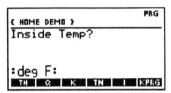
Figure 5-19

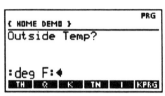
Figure 5-20

5   Hear tone and see data request for "Wall Thickness" (Figure 5-21). Key: .5, press ENTER.

6   Hear tone and see data request for "Wall Area" (Figure 5-22). Key: 200, press Enter.

Figure 5-21

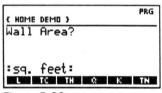
Figure 5-22

7   Hear tones, see solution for "k" (Figure 5-23, next page). Program automatically terminates

8   Local variable called "K" stores calculated value for "k". Other local variables are purged (Figure 5-24, next page).

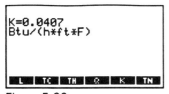
Figure 5-23

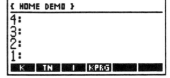
Figure 5-24

The Next Step

The program developed in this chapter was designed primarily to demonstrate program construction. It is a simple routine and provides only one calculation. There are many ways it could be improved. A more useful and interesting program, for example, would first ask the user to decide what variable is to be solved. The user then keys a code number in a table corresponding to, "Q", "k", "Th", "Tc", "L", or "A". Appropriate data would then be entered based on a "CASE-THEN-END" decision structure with subroutines housing corresponding equations and data requests. This is perhaps too complicated for a demonstration, but it is a good design for practical application. The challenge is to modify the program so that it better meets the needs for the calculation. Making such modifications will also be useful for learning programming.

Notes

Chapter 6

Basic Statistics

- Statistics for a Single Data Set
- Statistics Using Array functions
- Programs for Statistical Calculations

Notes

Chapter 6 - Basic Statistics

The HP-48G lends itself just as well to calculating basic statistics right from the keyboard, as it does when a program is written to control the operation. This chapter demonstrates both techniques. It also shows basic procedures for constructing a matrix table to analyze either for a single variable, or for data arrays.

Using the HP-48G for statistical operations is a two-step process. First, data are entered into a matrix table where they remain in RAM until they are cleared. Data are retained when the calculator is turned off and even when its batteries are changed. As the second step, required statistical operations are selected and executed from a number of built-in menus, or from a customized program.

Once data have been entered into the statistical table, it is an elementary procedure to edit them, decide the best way to display them, then calculate totals, averages, standard deviations, variances, standard errors, and other descriptive statistics for them. It is also easy in a separate statistical procedure to test mathematical models for entered data and define best-fitting equations for them. These are often difficult, complex and time consuming analyses, but not if someone knows how to use the HP-48G. Then, it's fast and easy.

A new subdirectory called, "STATS" will be used to demonstrate statistical operations. Keying RS, HOME exits whatever subdirectory was previously in use and returns operation to the HOME directory. Keying RS, MEMORY, NEW, ▼, α, α, S, T, A, T, S, ENTER, ✔CHK, OK, CANCEL, VAR, STATS constructs and directs operation to "STATS". The six blank rectangles at the bottom of the view window reassure there are no local variables in this newly constructed subdirectory.

> Create a new subdirectory for each new program, or application - see end of section to learn how to delete them.

Examples in this section use data displays with 2 digits to the right of the decimal in a FIX mode. Keying RS, MODES allows setting this display configuration. Step-by-step instructions throughout the section describe how to adjust view window displays to accommodate showing data in this mode. Other display styles would be generated, of course, in a similar way. None compromises the precision of a calculation.

There are several ways to calculate statistics and perform curve fitting operations. Only one technique is demonstrated in this chapter. Others are described in the book, "Mastering the HP-48G/GX". There is a lot to be said for reviewing several different ways of making a calculation, then deciding what's best for a particular operation. There are also advantages to developing one's own computational style.

The first example introduces techniques for accessing the built-in (ROM) statistical menus and matrix table for data entry and editing. As for other examples, only a few data are used so that procedures are seen more clearly. After reviewing the following example and executing the steps required for its solution, it's a good idea to generate new sets of data and try the same techniques for them. It all fits in place with a little practice.

Example: Statistics for a Single Data Set

(Note: Calculations are made directly from the keyboard using built-in statistics menus. A program at the end of this chapter shows how to make these calculations automatically.)

Problem Statement

Steady state temperature measurements were made at five sites in a viscous fluid presumed to be thermally uniform (Table 6-1). What is the average, standard deviation, variance, total, minimum and maximum values for this sample?

Table 6-1

Site	Temp. (F)
1	46.39
2	52.14
3	47.36
4	51.89
5	49.36

Chapter 6 - Basic Statistics   103

Solution

Step	Press
1	This solution begins with the blank display and empty local variable sites in {HOME STATS}, as shown in Figure 6-1. RS, STAT displays Figure 6-2.

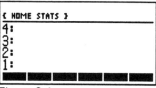
Figure 6-1

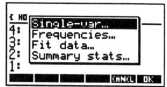
Figure 6-2

2	With "Single Var..." highlighted, press "OK" to display the menu for "SINGLE-VARIABLE STATISTICS" (Figure 6-3). Press "EDIT" to see the currently empty matrix table (Figure 6-4). The number at the lower left of the view window ("1-1") indicates the cursor is positioned by default at the first cell in the first column.

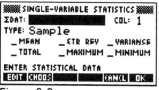
Figure 6-3

Figure 6-4

3	Key first data entry and see number appear at the lower left of the view window (Figure 6-5, next page). Press ENTER to place this number in the first cell of the matrix table (Figure 6-6, next page). Digits to the

104 Chapter 6 - Basic Statistics

Figure 6-5   Figure 6-6

right of the decimal are not shown, as indicated by the three dots. The cursor is automatically positioned to the first cell of the second column.

4   All digits are shown for data in the matrix table when "WID→" is pressed (Figure 6-7). Use the cursor control keys to highlight cell "2-1" (Figure 6-8).

Figure 6-7   Figure 6-8

5   Key next entry, press ENTER and repeat for all data. Cursor automatically goes to next cell in column (Figure 6-9, next page). When last entry has been made, press ENTER to return to "SINGLE-VARIABLE STATISTICS" menu (Figure 6-10, next page).

Chapter 6 - Basic Statistics

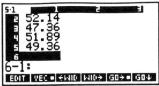
Figure 6-9

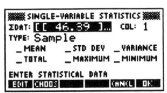
Figure 6-10

6   Move cursor to "___ MEAN" (Figure 6-11) and press "✔CHK" (Fig. 6-12;).

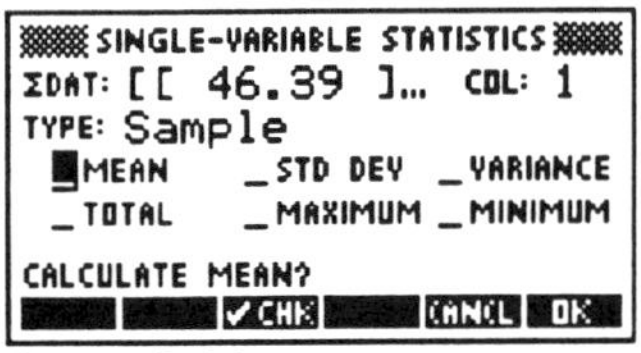

Figure 6-11

Figure 6-12

7   Select other statistics by highlighting the choice and pressing "✔CHK" (Figure 6-13). To make calculations, press "OK" (Figure 6-14).

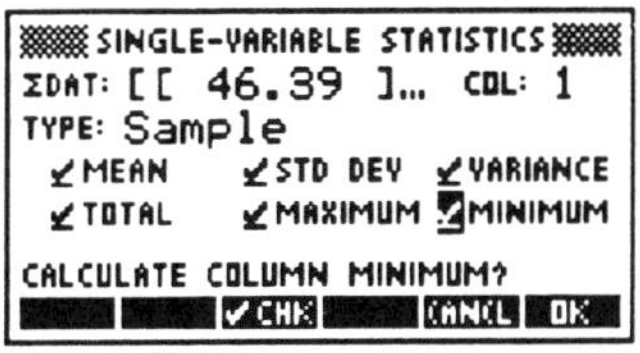

Figure 6-13

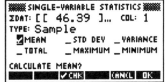
Figure 6-14

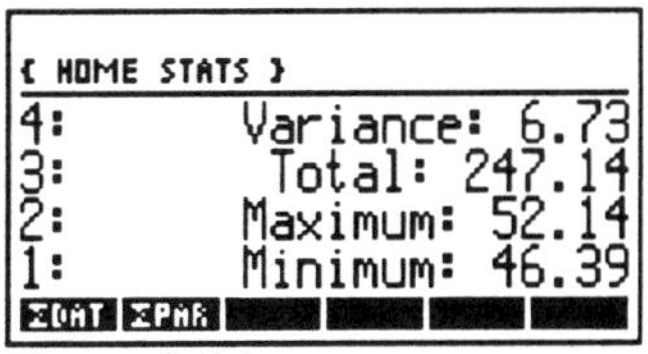

8   Press ▲ enough times to see solution for MEAN that is initially displayed off-screen at Line 6 of the view window (Figure 6-15). Using the ▲ or ▼ cursor control keys allows seeing all selected solutions. CANCEL returns the variable menu for {HOME STATS}. To remove the local variables created with the statistical operations, key: LS, { }, press "ΣDAT", "ΣPAR", ENTER, (Figure 6-16) then key: LS, PURG. To clear the STACK registers, key: LS, CLEAR. The {HOME STATS} subdirectory is now empty (Figure 6-1).

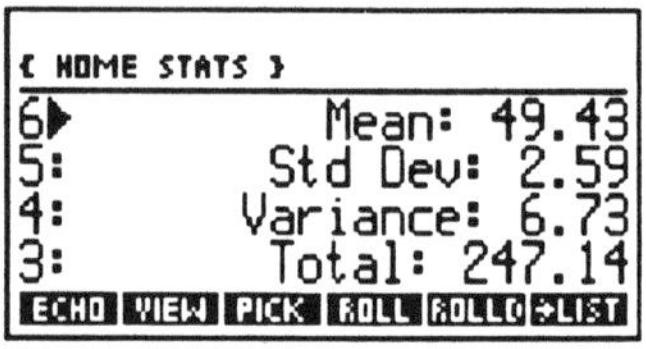

```
{ HOME STATS }
6▶         Mean: 49.43
5:       Std Dev:  2.59
4:       Variance: 6.73
3:          Total: 247.14
[ECHO][VIEW][PICK][ROLL][ROLLD][→LIST]
```
Figure 6-15

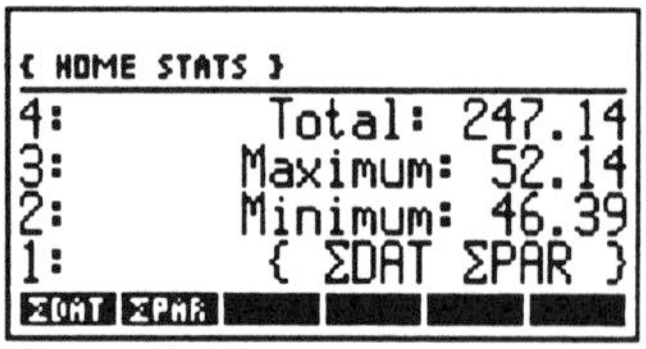

```
{ HOME STATS }
4:          Total: 247.14
3:        Maximum: 52.14
2:        Minimum: 46.39
1:           { ΣDAT ΣPAR }
[ΣDAT][ΣPAR][    ][    ][    ][    ]
```
Figure 6-16

Placing data in other cells in the matrix table to construct more than one column of entries provides several opportunities for additional statistical operations. It also sets the stage to determine mathematical relationships between dependent and independent variables by curve-fitting. The deduced equation that adequately defines these relationships can then be used to make predictions. These powerful analytical techniques are demonstrated next.

Example: Statistics Using Array Functions

Problem Statement

A chemist investigates the activity of a newly discovered enzyme, "EZ1" and its effects on a substrate, "SS1". Ten batches of test mixtures with different enzyme concentrations are measured for their final concentrations of "SS1" (Table 6-2).

Chapter 6 - Basic Statistics

Table 6-2: Data

Mix	EZ1 (mg/dl)	SS1 (mg/dl)
1	1.07	44.82
2	0.68	60.05
3	0.09	93.47
4	2.45	15.92
5	1.55	31.27
6	4.89	2.55
7	0.78	55.71
8	0.82	54.06
9	2.01	22.05
10	3.11	9.71

Part I:

What are the averages and standard deviations for the enzyme and substrate concentrations for the samples?

Part II:

Based on measured data, what is the "best-fit" equation that predicts how substrate concentration varies as a function of enzyme concentration?

Part III:

Using the "best-fit" model, what enzyme concentrations would produce substrate concentrations of 7.32, 10.01 and 50.00 mg/dl? What substrate concentrations would be produced by enzyme concentrations of 0.02, 0.20 and 3.76 mg/dl?

Chapter 6 - Basic Statistics

Solutions

Part I:

> *It is essential that all previously entered data be cleared from the statistical table when starting a new calculation.*

Step 1:

Calculations are made in the {HOME STATS} subdirectory. Key: RS, STAT. With "Single-var..." highlighted, press "OK". Had data not been cleared from RAM at the end of the last exercise, pressing NXT, then RESET, then ▼ to highlight "Reset all", then pressing "OK" would clear all data from the statistical table.

Step 2:

EZ1 concentration in this test is the independent variable, defined as "X". SS1 concentration is the dependent variable, defined as "Y". These designations follow statistical convention and have no necessary relationship to the X STACK REGISTER, or the Y STACK REGISTER, or data listed in them.

On the first "page" of the "SINGLE-VARIABLE STATISTICS" menu, press "EDIT". Key the enzyme concentration for the first mixture of "1.07", then press ENTER. Key the corresponding concentration of SS1, "44.82", then press ENTER. Press ▼ once to position the cursor at cell "2-1".

Key the enzyme concentration of ".68" for the second mixture, press ENTER. Key "60.05", then ENTER to list the corresponding SS1 concentration. Note that the cursor automatically positioned itself to cell "3-1" for the next data entry. Continue to enter enzyme and substrate concentrations, then press ENTER when all data are listed in the table.

As indicated by data arrangements in Table 6-2, it is unimportant that data for mixtures are entered in any other than random order. It is essential, however, that corresponding data for EZ1 and SS1 for any mixture are aligned in the same row.

Step 3:

In the "SINGLE-VARIABLE STATISTICS" menu, a notation at the upper right of the view window indicates that calculations will first be made on data in "COL.1". As data were ordered in this example, this is information about EZ1 concentrations. Use the cursor controls to highlight, then "✔CHK" categories for "___MEAN" and "___STD DEV.", then press "OK". Solutions appear at Lines 1 and 2 of the view window to show that the average EZ1 concentration across mixtures was "1.75" mg/dl and its standard deviation was "1.43" mg/dl.

To make the same calculations for SS1 concentration, key: RS STAT, with "Single-var..." highlighted, press OK, move the cursor to highlight "COL.1", key "2", and press ENTER. Calculations will now be made for data in column 2 of the statistical table. "✔CHK" categories for "___MEAN" and "___STD DEV.", then press "OK". Solutions appear at Lines 1 and 2 of the view window to show that the average SS1 concentration across mixtures was "38.96" mg/dl and its standard deviation was "27.92" mg/dl. LS, CLEAR removes all data from the STACK.

Part II

Step 1:

From the {HOME STATS} display, key: RS, STAT, press ▼ twice to highlight "Fit data...", then press "OK".

Step 2:

In the "FIT DATA" menu, press ▼ once to highlight the "X-COL:" position. Key: "1" and press ENTER. This defines that the equation to be derived is based on data for the independent variable (EZ1) in "COL.1". The corresponding dependent variable (SS1) shows as the "Y-COL:2".

Press ▼ once to highlight the selection for the "MODEL:". Press "CHOOS", then move the cursor to the bottom of the view window to highlight "Best Fit", then press "OK". Pressing "OK" again displays solutions in the view window at Line 1 for covariance ("-1.54"), Line 2 for the correlation coefficient ("-1.00"), and Line 3 for the "best-fit" equation ('99.99*EXP(-0.75*X...); to see the full equation, press the erase key (left pointing arrow) twice.

The negative sign for the calculations of covariance and the correlation coefficient reveal that substrate concentration is inversely proportional to enzyme concentration. The "best-fit equation" indicates that

$$substrate\ concentration = 99.99\ e^{-0.75(enzyme\ concentration)}$$

when both substrate and enzyme concentrations have units of "mg/dl".

Part III

From the {HOME STATS} display, key: RS, STAT, press ▼ twice to highlight "Fit data...", then press "OK". Press "PRED". Press ▼ three times, then RC once to highlight "Y:". Key: "7.32", then ENTER to enter the substrate concentration, then move the cursor to highlight "X:" press "PRED", then "EDIT".

This calculation indicates that according to the "best-fit" model based on empirical data, an enzyme concentration of 3.484647046 mg/dl is required to yield a substrate concentration of 7.32 mg/dl. For the next calculation, press "OK", highlight "Y:", key "10.01", press ENTER, move the cursor to "X:" and press "PRED".

A substrate concentration of 10.01 mg/dl would be roduced for an enzyme concentration of 3.06751... mg/dl, and it would be 50 mg/dl for a mixture containing 0.923758... mg/dl of enzyme.

Substrate concentrations are predicted in a similar way by entering successively values of 0.02, 0.20 and 3.76 for "X:" and using "PRED" to predict corresponding values of substrate concentration "98.5047", "86.0606" and "5.95371" mg/dl, respectively.

Programs for Statistical Calculations

For many people, calculations of basic statistics are so commonly a part of their day-to-day professional lives, that having a program in the HP-48G to do them is more convenient than depending on keyboard operations.

The solutions made by keyboard operation or by a program are, of course, the same, but less time is invested in entering data and having to remember the next step for a complicated series. One of the major advantages of a program is, of course, that it routinizes operations. These techniques are demonstrated next.

Program Listing:

```
SUM
«CLΣ TN "decimal?"
":digits to right:"
INPUT OBJ→ FIX
CLEAR REQ »
```

```
REQ
«TN1 "Entry?"
":N:" INPUT OBJ→
'N' STO
 IF 'N≠0'
 THEN N Σ+ REQ
 ELSE TN TN CLLCD
 TOT "Total= " SWAP
 + 2 DISP MEAN
 "Mean= " SWAP + 3
 DISP SDEV DUP 'SD'
```

```
STO "SDev= " SWAP +
4 DISP NΣ 'T' STO
'SD/√(T-1)' →NUM
"SErr= " SWAP + 5
DISP 0 FIX NΣ
"N = " SWAP + 6
DISP 0 WAIT CLEAR {
SD T N ΣDAT } PURGE
2 FIX
 END»
```

```
TN
«2000 .1 BEEP»
```

```
TN1
«1500 .1 BEEP»
```

112   Chapter 6 - Basic Statistics

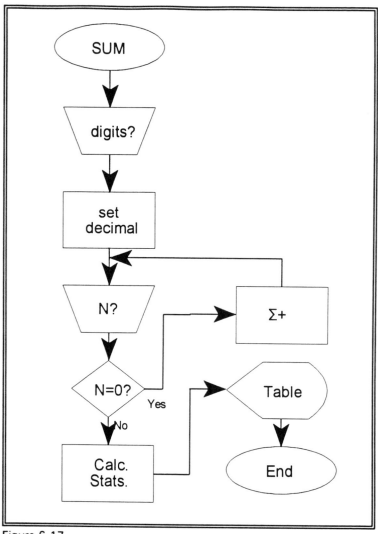

Figure 6-17

General Program Description

The SUM program (Figure 6-17) is stored in a subdirectory "SUMS". It calculates the total, average, standard deviation, standard error and tabulates the number of entries of an indefinite series of entries (data entries are: "19.30, 5.60, 8.93, 12.11 and 16.73" for this example), displaying results in a table.

When the program is initialized by pressing SUM, the user is first asked for the number of digits required to be shown to the right of the decimal. ("2" is used in this example; Figure 6-18). After ENTER is pressed, the user is then asked for the first number in the series ("19.30" for this example; Figure 6-19).

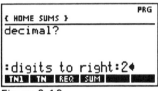
Figure 6-18

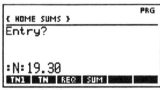
Figure 6-19

Other data are entered in a similar way. Entering a zero signals the end of the series and displays a table of solutions (Figure 6-20). Pressing any key clears the display.

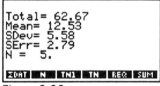
Figure 6-20

Detailed Program Description

The first instruction in the SUM program (CLΣ; see "Program Listing") clears data from the statistical table. Following a branch to a subroutine to activate a tone (TN), a prompt places "decimal?" at the top of the view window and ":digits to right:" at the bottom. Whatever number is keyed and ENTERed controls the FIX operation. The STACK is then cleared and the program is directed to another program list, REQ in which data are requested for calculation.

Chapter 6 - Basic Statistics

A different tone (TN1) is constructed, and the user is asked for a data entry with "Entry?" at the top of the view window and ":N:" at the bottom. Whatever number is entered is first stored in a local variable (N), then tested to see if it is a zero. An IF, THEN, ELSE, END construction branches the program to store the number in "N" if it is not zero, or begin the construction of the results display if it is zero. If the number is not zero, it is recalled from the local variable "N" and placed in the X STACK REGISTER by the instruction N. Σ+ stores it as an entry in the statistical table, and the program returns to REQ to request the next entry.

When a zero is entered in request to a data entry, two tones are sounded (TN TN), the display is cleared (CLLCD) and the ROM instruction, TOT, is used to calculate the total of entries in the statistical table and the phrase "Total = " is constructed. A space after the " = " symbol makes it easier to read results in the display. The STACK positions of the alpha phrase (now in the X STACK REGISTER) and the total of entries (now in the Y STACK REGISTER) are exchanged by SWAP and displayed together (+) at line 2 of the display by "2 DISP".

The average of data entries is calculated (MEAN) and displayed in a similar way at line 3 of the display. Standard deviation is calculated by SDEV, but duplicated (DUP) in the STACK to support both the display of standard deviation, but also the calculation of standard error which is not included in the HP-48G's ROM. The number generated by SDEV is stored as a local variable (SD) and displayed at line 4 of the display with the phrase, "Sdev= ".

The total of entries is determined by a ROM instruction, "NΣ", and stored in a local variable "T" to represent a total. Standard error is calculated as the ratio between the standard deviation and the square root of the total minus 1 by the phrase, "(SD/√(T-1)" and displayed with an appropriate alpha construction ("Serr= ") at line 5 of the view window.

The display is then adjusted to display no digits to the right of the decimal (0 FIX) to prepare for the display indicating the total number of entries. Because these data are a count, a display as a whole number is appropriate. NΣ defines the count which is displayed with the phrase "N = " at line 6 of the view window. The instruction "0 WAIT" holds the display indefinitely until any key is pressed. This construction allows the user to read and copy results with no time restriction. When a key is pressed, local variables generated by the program are purged by "{SD T N ΣDAT}", the display of digits is arbitrarily set to 2 by "2 FIX" and the program ends.

The Next Step

Like any customized program, the SUM program was designed to do a specific job. With a few minutes of familiarization it becomes a practical tool for making basic calculations. Changing the program, however, might make it more useful in some applications.

For example, using a zero to terminate data entry may be impractical if, in fact, a zero is a legitimate data entry in a series. One solution is to use the program as it is now constructed an enter a non-zero number, but one that is so small, it doesn't affect the precision of the calculations of total, standard deviation, mean or standard error. The number, "0.00001" would have served this purpose for data series used in the example.

Another solution is to rewrite the program so that an alpha data entry, not a numerical one, controls the IF, THEN, ELSE END conditional test. An appropriate, short explanation on an opening display would explain this strategy to the user. The basic structure of the SUM program supports adding additional calculations, redefining the calculation for standard error, saving results for subsequent calculations, displaying data as a bar graph, or making other stylistic changes.

Deleting Local Variables and Subdirectories

The HP-48G doesn't have to be used very long before the HOME directory becomes loaded with many subdirectories, each of which may have its own series of subdirectories or local variables. In the beginning, this doesn't usually encroach on available memory, but it does add to directory clutter and sometimes makes it hard to find needed programs.

In the same way that it is good practice to generate a new subdirectory for each application, it is also worthwhile to eliminate local variables and subdirectories when they are no longer needed. Both procedures are easy, but must be done with care. There is no way to recover from an erroneous erasure.

Local variables in the HOME directory, or in a subdirectory, are displayed by pressing VAR. Each is identified by its name shown in one of the small rectangles at the bottom of the view window. There are typically many of them. Those that extend to additional "pages" are viewed by pressing NXT.

To eliminate a single local variable, first press the apostrophe key, then the white key under the view window associated with the local variable. Pressing LS, PURG removes the local variable. To eliminate more than one local variable at a time, key: LS, { }, then press in any sequence the white key corresponding to each targeted local variable. Press ENTER when all local variables have been identified. Pressing LS, PURG removes the defined group of local variables.

Subdirectories are distinguished from local variables by the small tag that extends upward from the left of a rectangle at the bottom of the view window. Once a subdirectory has been purged of its local variables, it is itself erased by pressing: RS, MEMORY, (use the ▼ and ▲ cursor control keys to highlight the subdirectory), NXT, PURG. Pressing CANCEL displays the view window which no longer lists the eliminated subdirectory. Purging a subdirectory using functions accessed by RS, MEMORY requires that the targeted subdirectory is empty of local variables.

As an alternative, a subdirectory can be eliminated even though it still contains local variables. The procedure deserves special care. Erasing the wrong subdirectory and its local variables will not only

Chapter 6 - Basic Statistics

delete possibly valuable data, it will also cost the hours of having to re-key one or more possibly long programs. To eliminate a subdirectory and the local variables it contains, press the apostrophe key, then press the white key associated with the subdirectory, and finally, press ENTER, LS, MEMORY, DIR, PGDIR. Pressing VAR shows the subdirectory is gone.

A program that will automatically calculate curve-fitting statistics, along with many other programs, is described in, "Mastering the HP-48G/GX"; call: 1-800-228-0810.

Notes

Chapter 7

A Practical Problem
Heat Stress in the Workplace

- The Problem and its Background
- Physical Avenues of Heat Exchange
- Physiological Factors
- The Human "Heat Stress Index"
- Developing the HP-48G Program
- Sample Problems
- Complete Program List and Description

Notes

Chapter 7 - A Practical Problem

The Problem

Professionals in many different fields commonly need to solve equations, or sets of them, either symbolically, or numerically, to make decisions. If calculations are complicated, or if there are lots of them, many people have to depend just on the computer back at the office or on the one at home. Because these machines are not portable and sometimes others are not accessible elsewhere, decisions are often delayed. In the worst case, they are not made at all because of the inconvenience.

Such has been the case for the set of equations that calculates a human "Heat Stress Index" (HSI) and the biological heat transfer and storage information related to it. These calculations define how factors of conductive, radiative, convective, and evaporative heat transfer interact with those of metabolic heat production when someone is heat exposed.

The HSI equations were developed more than fifty years ago. They provide valuable first-order insight into the potential dangers of heat stress on-the-job and yield many useful guidelines for managing related environmental and personal conditions. But, calculations using the equations are not often made because of the inordinate amount of time and effort required to solve them by hand. The power of the HP-48G goes a long way in making HSI computations practical, as it does in other similar applications.

The purpose of this chapter is to show how the HP-48G is valuable as a portable and powerful computer for handling a set of complex calculations with accuracy and ease. The chapter demonstrates how once a program is designed to request data, solving associated equations to get valuable information about a complex system is fast, accurate and no more than a few button-pushes away.

Once the program for calculating the equations for the "Human Heat Stress Index" is developed, it is also useful for making series of "what-if" calculations, as demonstrated by two examples at the end of this chapter. Testing the effects of small changes in the values for each of the independent variables is an excellent way to learn first-hand what is most important and how all the complex factors interact. Because the HP-48G allows such calculations to be made rapidly and easily, it is an excellent tool for this kind of learning.

Not everyone understands the complexity of body temperature regulation. Also, not all appreciate that body heat content can be allowed to vary considerably during periods of heat and cold stress, but deep body temperature must remain in a narrowly controlled range. These insights are important for using the HSI program and interpreting the results of its calculations. The next section provides basic information about human body temperature regulation and heat stress.

Background

Why Some Animals Regulate Body Temperature

Unique among life forms, birds and mammals, including people, share the important characteristic of regulating body temperature to be in a narrow range. Other animals, like fish, reptiles, and insects, also depend on a near-constant body temperature, but they achieve it more through behavioral than physiological means. Having a built-in control system gives considerable freedom from the capriciousness of the climate and is of survival importance.

The advantage to all animals is that with a stable internal temperature between about 96F and 99F (35.6C to 37.2C), metabolic processes operate with constant and high efficiency. The upper limit of this range, however, is dangerously close to that at which cells die. Sustained human body temperature much above about 105F (40.6C) may produce tissue damage and death. Body temperatures below about 95F (35C) induce lethargy, mental confusion and seriously compromise function.

ue, of course, has other importance. It delivers oxygen
to support metabolic functions, and it removes carbon
metabolic waste products from it. Even so, it is the major
tion system of the body.

mon knowledge and experience that fingers and toes, as
ds and feet, decrease their heat content and drop
uring cold exposure. This is because thermoregulatory
ease blood flow and convective heat transfer to these
limit body heat loss. This comes about by constricting
od vessels, reducing their net cross-sectional area and
r resistance to blood flow.

the temperature of peripheral tissue also lessens the
between the skin surface and the environment. The
educe body heat loss. Also, additional time is gained
body temperature constant by draining heat from
heral tissues.

Heat Exchange at The Body Surface

eat transfer occurs at the body surface when a fluid
ater) flows over it. If it is warmer than average skin
ut 95F (35C)), it heats the body, and *vice versa*
transfer depends on the convective heat transfer
ea of exposed skin and the temperature difference
and the skin. If there is no thermal gradien
surface and the fluid flowing over it, then there i
transfer, despite fluid flow.

perature is less than average skin temperature
over the skin surface is an effective way t
t. This is why, in part, removing clothing an
s of the skin to rapidly flowing, cooler air stirre
breezes are so effective for body cooling an
mmer day, especially after a period of exercise

perature is far enough below average ski
a danger of body cooling below a normal leve

How Body Temperature is Regulated

A nearly constant and safe internal body temperature is achieved by a neurophysiologically-driven control system. Thermal inputs are collected at millisecond intervals from hundreds of thousands of sites in the skin and from deep body tissue. A controller no larger than a pea at the base of the brain then manages how and where heat is distributed in the body, adjusts effective thermal conductivities of tissues, regulates the volume flow of sweat to the skin surface for the dissipation of heat by evaporation, and controls body heat production by metabolism.

This controller must operate continuously against the challenges of environmental heat and cold exposure, and protect against excessive heat gain during exercise. Despite continuous operation under a constant load, the control system for body temperature regulation normally works well for a lifetime. It makes the required minute-to-minute adjustments in heat distribution, storage and dissipation so effectively, that most people are unaware of its operation, even at times of peak demand.

As for any other thermoregulated system, a constant level of body heat content and temperature depends on dynamically balancing the rate at which the body gains heat and the rate at which it loses it. Although transients in heat storage induce temporary imbalances, net body heat exchange with the environment over a long period must be zero. Even during phasic heating and cooling, stored body heat must be distributed so that organs deep in the body, especially the brain, remain as thermally stable as possible, with no more than a degree or so variation around 98.6F (37C).

Most animals, including humans, buffer changes in deep body temperature by using peripheral tissue as a thermal shock absorber. During exercise, for example, contracting muscles increase heat production by as much as seven times the resting level. This adds heat to the body at rates near 700 Watts (about 2400 BTU/hour). Whatever heat cannot be dissipated to the environment is stored in tissue of the body's periphery. Deep body temperature is protected until this heat storage capacity is filled. Similarly, during cold exposure, deep body temperature is protected until heat stored in peripheral tissue is drained.

About 50% of the body mass serves as such a thermal capacitor. Primary functions of the body's temperature regulating system are to control the distribution of heat in the body and regulate its exchange with the environment. The net effect is that heat in the body (so-called, "total body heat content") varies greatly, whereas deep body temperature normally remains stable.

The Physical Avenues of Heat Exchange

Accurately interpreting calculations of the human "Heat Stress Index" depends on understanding how the body interacts with the physical avenues of heat transfer, and how it deals with heat produced by metabolism.

Heat Exchange by Thermal Conduction

As for an inanimate object, heat is exchanged in the living body by thermal conduction between objects that are in contact with one another which have different temperatures. Inside the body, organs and tissues at different temperatures exchange heat across their areas of contact by conduction. Outside the body, skin exchanges heat with objects touching it, as for example, when feet are in contact with the cold (or hot) ground. There is, of course, no net heat transfer by conduction if objects have the same temperature at the area of their contact.

Because humans typically have only small areas of skin in contact with masses around them, heat exchange by conduction is usually unimportant in overall thermal balance. For example, less than 1% of an adult person's about 21.5 square feet (2 square meters) of skin is typically in contact with the ground, and even that is usually covered with material offering some degree of thermal insulation. Heat exchange by thermal conduction is ignored in the calculation of the human "Heat Stress Index".

Heat Exchange by Convection

Convective heat transfer comes fr
fluid over a surface that is at a differe
only inside the body, but also betw
surrounding environment. Convective
at both sites for the regulation of bo

Convective Heat Exchange Insic

Convective heat transfer occu
movement of blood. This is by f
internal heat distribution and exchar
from the contracting ventricles of
resting person at a rate of abou
liters/minute). It flows through a
a total cross-sectional area of a
microscopically intimate contac
output may increase to about 35
during exercise.

Such large volume flow ra
finely dividing channels with a l
heat transfer and distribution,
small. There is no convective
mass flow of blood, if the
between the blood and the t

Convective heat transfe
loss, depending on the c
gradient. Blood that is c
removes heat when it flow
by convection through
metabolically active orga
cooler tissues, and deli
environment.

Without such conv
muscle would rise to let

flow to tiss
and nutrient
dioxide and
heat distribu

It is com
well as han
temperature
reflexes decr
extremities to
peripheral blo
increasing the

Decreasing
thermal gradien
net effect is to
in keeping dee
storage in perip

Convective

Convective h
(typically air or w
temperature (abc
The rate of heat
coefficient, the ar
between the flui
between the skin
no convective hea

When air tem
increasing air flov
dissipate body he
exposing large area
by fans or natural
pleasant on a hot, s

When air tem
temperature there is

decreasing convective heat transfer at the skin surface is the essential first-line of defense. This is why, in part, that covering large areas of skin with thermal insulation, especially that with even a thin, but densely woven cover, serves as a windbreak and reduces convective heat loss. All thermal insulation at the body surface has a similar function. It reduces convective heat transfer to the environment. This is as true for a pair of woolen gloves, a sweater or a hat, as it is for a diver's insulated wet-suit. Any material that traps a layer of still air at the skin surface is a good thermal insulator.

Heat Exchange by Thermal Radiation

All objects above zero K constantly radiate thermal energy. Whether there is net exchange between the facing surfaces of two (or more) objects depends on their differences in temperature, emissivity, and exposed areas. Net thermal radiative heat transfer is zero, of course, if there is no thermal gradient. For example, there is no net transfer of radiant heat inside the body, although there are surfaces that face one another across an air gap, as do the inner surfaces of the trachea, bronchioles and lungs. Because these structures are deep in the body, they are all at virtually the same temperature.

Thermal radiation, however, can be a major avenue for heat transfer at the body surface. High thermal radiant gains come from direct exposure to the sun, of course, and from the surfaces of hot machinery, surrounding buildings, or other objects in the near environment that are warmer than the skin or the surface of clothing.

There is radiant heat loss from the body when skin or clothing is exposed to the night sky, to cold walls or to other objects whose surfaces are colder than that of clothing or skin. It is a common experience to be exposed to many different objects simultaneously, some of which serve as radiant heat sources and others that are radiant heat sinks. Net radiant exchange is the algebraic sum of their effects.

There are many examples of how people are affected by radiant heat transfer. Fire-fighters, for example, often wear protective clothing with a brightly polished outer layer whose emissivity is so low, it effectively reflects thermal radiation. Also, most people know from direct experience that sitting in a room with a poorly insulated exterior wall, whose inner surface is cold, soon leads to feeling chilled, even though air temperature is in a comfortable range.

Many animals other than humans know about the rules of radiative heat transfer and how to bend them to their advantage, even though they don't have an HP-48G. Camels, for example, huddle in groups even during the hottest desert days. Must humans know how huddling helps conserve body heat when it's cold, but camels know how to reduce heat gain by grouping together when it's hot. In a first consideration, huddling on a hot day seems counterintuitive for keeping cool (not getting hot). It would be for humans because we depend on heat loss by the evaporation of water at the skin surface. It works well, though, for the camel.

A camel by itself gains radiant heat not only from the direct rays of the sun, but also from surrounding sand, buildings and other objects that have temperatures greater than that of its skin. A camel in the center of a huddled group, however, has no net radiant heat gain at skin surfaces that face those of other animals, because their skin surface temperatures are all about the same. Reducing radiant heat gain in this way at its flanks, head and rear gives a substantial survival advantage. These body areas account for a sizeable portion of the animal's total body surface area. An animal at the edge of the group has less of an advantage, but still gains radiant heat from only surfaces that are not facing other animals.

Although the camel's huddling behavior seems to be a marginally effective thermal defense, it provides many secondary advantages. Whatever heat an animal doesn't gain by protecting itself in this way, is heat it doesn't have to dissipate by the evaporation of water - a precious commodity in the desert. Also, huddling not only reduces heat gain by thermal radiation, it limits the flow of hot desert air over the skin to reduce convective heat gain, and gives a good opportunity to rest, thereby reducing metabolic heat production. Such a simple adaptation as huddling has, in fact, many far-reaching advantages.

Heat Loss By Evaporation

Evaporation is only an avenue for heat loss. It always requires heat for water and other liquids to go to a vapor state. The heat is drawn from the surface where the process takes place. It is released, of course, when the material goes from a vapor to a liquid state by condensation. Heat gain by condensation is seldom encountered in human heat stress, and the effect is ignored in the calculation of the human "Heat Stress Index". Heat loss by evaporation, however, is of major consequence.

Heat loss by the evaporation of water depends only secondarily on temperature. It is a primary function of the difference in water vapor pressure between the surface of the liquid and that of the surrounding environment. If this vapor pressure gradient is zero, then there is no evaporation and no evaporative heat loss. For this reason, contrary to popular opinion, no heat is lost by the process of sweating. Heat is lost only when the water in sweat evaporates.

If there is no evaporation, sweating only succeeds in removing much needed water and electrolytes from the body. Also contrary to expectation, sweat dripping from the skin doesn't cool. Although heat is lost, so is a matching loss of mass, and there is no net decrease in skin surface or body temperature. The water in sweat must evaporate in order to cool.

Sweat production and evaporative cooling are often overlooked, even though many quarts, even gallons, of water may be lost on a hot day, especially with exercise or prolonged physical work. Water vapor is invisible - it leaves the body surface and carries away heat without notice. Each of the 5 million or so microscopically small sweat glands secretes water sometimes at rates of only a few microliters per minute. Total sweat production, however, may peak for short periods at rates of about 3 quarts per hour (about 50 cc per minute), and as much as 12 quarts (11.4 liters) can be lost in a 24 hour period.

As environmental temperatures approach 95F (average skin surface temperature), evaporation emerges as the only avenue for heat loss. Because of narrowing thermal gradients, thermal conduction, radiation and convection no longer dissipate body heat. When environmental temperature is greater than 95F, they become avenues of heat gain.

Metabolism is, of course, always a source of heat gain. Under these conditions, sweat glands play an essential role in maintaining life and are as vital body organs as are the heart, lungs and kidneys. Death may be no more than minutes away if sweat glands fail. "Sweat gland fatigue" is extremely rare, however, and these glands usually continue to carry successfully the burden of thermoregulation. How fast heat can be lost by skin surface evaporation, and how fast it must be lost for a particular exposure, are important calculations in the human "Heat Stress Index".

Physiological Factors

Table 7-1: Heat Production for Different Physical Activities		
Activity	Watts	BTU/hr
Sleeping	50 to 80	250 to 300
Sitting: - quietly - moderate arm movements - moderate arm and leg movements	125 150 200	400 500 650
Standing: - light work - moderate work	250 350	750 1000
Pick and shovel work	650	2000
Hardest sustained activity	700	2400

Metabolic Heat Production

Data in Table 7-1 show representative heat production rates for different human activities. Even when someone is asleep, heat is produced as a metabolic byproduct. Prior to the first quarter of this century, many considered that the "animal heat" produced by living organisms was somehow different from that coming from inanimate sources. But it is now universally recognized that "calories are calories are calories ...". Heat has the same basic physical properties no matter what its source.

Heat from physical activity kills many otherwise healthy, young people each year, even when ambient temperatures and conditions are mild. Sustained high levels of physical activity, as among marathon runners and heavily exercising athletes, may produce heat at rates too high to be dissipated. Body heat content progressively rises to produce either "exercise-induced heat exhaustion" and collapse, or "exercise-induced heatstroke" and death. Thermal strain is often intensified because of an athlete's padded clothing and headgear which thermally insulate.

Human Variations in Heat Strain

Thermal stress is defined by the characteristics of the environment, like ambient temperature, humidity, air flow, radiant heat exchange and other factors. It is the same for everyone in the same environment. *Thermal strain* is defined by the cost to the individual in facing the thermal stress. It is different for everyone, because each person has unique characteristics for withstanding heat or cold exposures. There is no way to reliably predict a person's level of thermal strain based solely on knowing the thermal stress.

People vary greatly in how they tolerate thermal stress. What is a comfortable and safe heat (or cold) exposure for one person, may be lethal for another, even when they are dressed similarly and are performing the same physical activity. There are many factors affecting thermal strain. Some are age, physical fitness, hydration state, amount and distribution of body fat, state of health, experience, and history of prior exposures. Body water-and-electrolyte and acid-base balances, cardiovascular condition, disease history, medication plans, training and many other factors also mitigate safety and comfort in any thermal stress, even for the same person from time-to-time.

For these reasons, the human "Heat Stress Index" does not provide safe predictions about how well a particular person will tolerate a heat stress. Its calculations are, nonetheless, useful. They give insight into the net interactions of metabolic heat production and the physical avenues of heat exchange and transfer.

When the HP-48G is used to make the calculations, it is informative, for example, to see how small increases in air velocity greatly increase convective and evaporative heat loss. The "Heat Stress Index" gives important information about thermal stress, but not about thermal strain.

All calculations associated with the "Heat Stress Index", especially the one that predicts a "safe exposure time", must be interpreted in the context of the artificially constructed hypothetical problem. They can never be applied with any safety or validity to the circumstances facing a particular person, no matter how well defined are the physical and metabolic factors.

The Human "Heat Stress Index"

General Description

The "Heat Stress Index" (Table 7-2) is a dimensionless number that indicates on a scale from zero (no heat stress) to one hundred (the maximum tolerable heat stress) the severity of a thermal exposure. Calculations evaluate the environmental factors related to relative humidity, convective and radiative heat gains, and the human factors

Table 7-2: The *Heat Stress Index* (HSI)

HSI	Heat Strain
100	Maximum tolerated
90	very severe
80	
70	
60	severe
50	
40	
30	mild
20	
10	
0	none

of sweating and metabolic heat production.

When the "Heat Stress Index" is calculated to be greater than one hundred, a "safe exposure time" is predicted. This indicates the length of time a normal, healthy, young person could sustain the exposure and still keep internal body temperature in a non-damaging range.

Using the program

The logic flow diagram for the HP-48G HSI program is shown in Figure 7-1 (next page). The program listing is at the end of the chapter. Screen displays for each stage of the program are shown in Figures 7-2 to 7-15 in association with calculations for "Sample Problem No. 1: A Hot Office."

As shown in Figure 7-1, once the program is started, the user is first asked to designate whether English or Metric units will be used in calculations. Unit selection must remain the same throughout program use. The user then provides data with appropriate units for dry bulb temperature (TDB), wet bulb temperature (TWB), globe temperature (TGT), air velocity (V), and the person's work rate (W).

Once all data have been entered, the program calculates and displays the "Heat Stress Index" and a "safe exposure time". Pressing any key advances the program and displays results of the next calculation, heat transfer by thermal radiation and the percentage of the total heat load it represents. Pressing any key shows results of the next calculation for convective heat transfer, then the metabolic heat load.

Final calculations show ambient relative humidity, the evaporative heat loss rate required for thermal stability, and an estimated maximum evaporative heat loss rate in consideration of physical and physiological limitations. To make "what-if" calculations easy, as a last step, a display asks if the program is to be run again with new data. If the answer is "Yes", units are redesignated and new data entered. If the answer is "No", the program terminates.

Chapter 7 - A Practical Problem

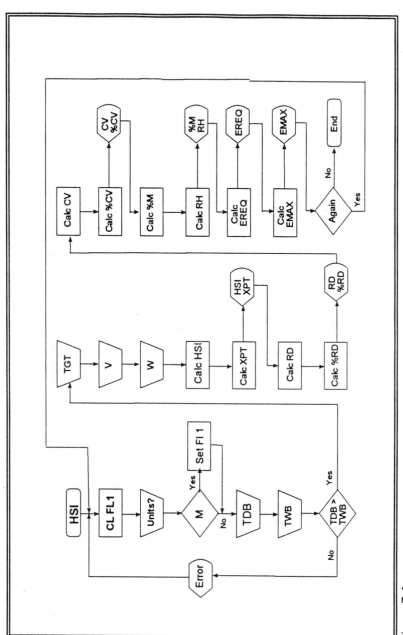

Figure 7-1

Symbol Definitions

CV = convective heat exchange (BTU/hr or Watts)
E = English units (e.g., deg. F; BTU/hr)
EMAX = maximum evaporative heat loss (BTU/hr or Watts)
EREQ = required evaporative heat loss (BTU/hr or Watts)
HSI = heat stress index (dimensionless)
M = Metric units (e.g., deg. C; Watts)
MRT = mean radiant temperature (deg. F or C)
PDB = pressure of saturated water vapor at TDB (mbars)
PH2O = water vapor pressure (mbars)
PV = water vapor pressure (torr)
PWB = pressure of saturated water vapor at TWB (mbars)
RD = radiative heat exchange rate (BTU/hr or Watts)
RH = ambient relative humidity (percent)
TDB = dry bulb temperature (deg. F or C)
TDBC = dry bulb temperature (deg. F or C)
TGT = globe temperature (deg. F or C)
TWB = wet bulb temperature (deg. F or C)
TWBC = wet bulb temperature (deg. F or C)
V = air velocity (ft/min or m/sec)
W = metabolic heat production rate (BTU/hr or Watts)
XPT = "safe" exposure time (hr.min)
%RD = percent EREQ of infrared radiative heat transfer
%CV = percent EREQ of convective heat transfer
%MET = percent EREQ of metabolic heat production

Equations

HSI = 100(EREQ/EMAX)
XPT = 250(EREQ-EMAX)
RD = 15(MRT-95)
 = 15(TGT + 0.13($V^{.5}$)(TGT-TDB))-95
%RD = 100(RD/EREQ))
CV = 0.65($V^{.6}$)(TDB-95)
%CV = 100(CV/EREQ)
%MET = 100(W/EREQ)
%RH = 100(PH2O/PDB)
EREQ = W ± RD ± CV

$$= W + 15(TGT + (0.13(V^{.5}))(TGT-TDB)-95)$$
$$+ 0.65(V^{.6})(TDB-95)$$
$$EMAX = 2.4(V^{.6}(42-PV)$$
$$= 2.4(V^{.6}(42-0.75(PH2O))$$
$$PH2O = PWB-0.6748(TDBC-TWBC)$$
$$PWB = 1000(ANTILOG(28.59$$
$$-8.2(LOG(TWBC+273))+0.00248(TWBC$$
$$+273)-3142/(TWBC+273))$$
$$PDB = 1000(ANTILOG(28.59-8.2(LOG(TDBC$$
$$+273))+0.00248(TDBC+273)$$
$$-3142/(TDBC+273)))$$
$$MRT = TGT+0.13(V^{.5})(TGT-TDB)$$
$$PV = 0.75(PH2O)$$

Sample Problems

Sample Problem No. 1 (Table 7-3, next page) presents the circumstance of a mild heat stress. Imagine George works in an office that is not air-conditioned. With incandescent lights in the room, several computers operating, and with the copier nearby, it's a noticeably uncomfortable place to work. George sits at his desk most of the time making paper-and-pencil notations and using his computer. His estimated metabolic heat production for these tasks is about 500 BTU/hr (147 Watts).

George's office is almost unbearably hot one summer afternoon. The humidity is low, but everyone is complaining about the heat. The C.E.O. of George's company says he cannot buy an air-conditioner for the office. It would belabor his already heavy payments for the Mercedes and the yacht. Not to be defeated by it all, George borrows some test equipment and measures the "Before" data shown in Table 7-3 in order to make calculations with the "Heat Stress Index" program.

Once the program is started (Figure 7-2) and units designated (Figure 7-3), data are entered for the conditions of the office (Figures 7-4 to 7-7) and George's work rate (Figure 7-8). Figures 7-9 to 7-15 provide valuable insight into the heat stress of the office.

Table 7-3: Sample Problem No. 1 - A Hot Office

Measurement	Before		After	
	English	Metric	English	Metric
Dry Bulb Temp.[1]	85.0	29.4	85.0	29.4
Wet Bulb Temp.[1]	65.0	18.3	65.0	18.3
Globe Temp.[1]	86.0	30.0	86.0	30.0
Air Velocity[2]	30.0	.15	800	4.1
Work Rate[3]	500	146	500	146
Calculations				
HSI	55.4	55.7	2.6	2.5
Exposure Time[4]	indef.	indef.	indef.	indef.
Radiation[3]	-124.3	-36.2	-79.8	-22.0
% Radiation	-38.2	-38.0	-130.0	-125.7
Convection[3]	-50.0	-14.6	-358.7	-106.5
% Convection	-15.4	-15.4	-584.0	-607.7
% Metabolism[3]	153.5	153.4	814.0	833.5
% Rel. Humidity	33.0	33.0	33.0	33.0
Evap. Required[3]	325.7	95.2	61.4	17.5
Evap. Maximum[3]	588.2	170.8	2400	702.9

Units: Note	English	Metric
1	deg. F	deg. C
2	ft/min	m/sec
3	BTU/hr	Watts
4	Hr. min	Hr. min

Slight differences between corresponding values are because of rounding. Calculations are for a hypothetical case only and are not intended to be applied in a real-life situation.

Chapter 7 - A Practical Problem

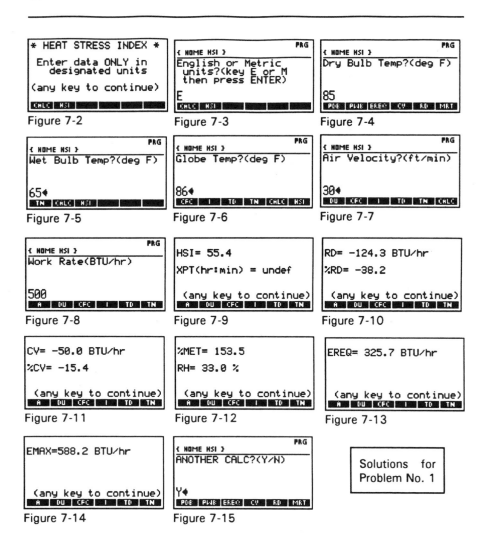

Solutions for Problem No. 1

George's first guess was right. The humidity is low, but the air temperature is high, as most of the office workers already knew. The calculated "Heat Stress Index" shows the conditions of the room to be survivable, but they certainly aren't pleasant. Although radiant heat is low and the work rate can't be reduced, air velocity in the room is that of still air. No wonder convective heat loss rates are so low (50 BTU/hr). George buys a couple of fans.

Data for "After" conditions in Table 7-3 show that increasing air flow with the fans was just enough to remediate the heat stress for the office and drastically reduce the "Heat Stress Index". There are two primary effects. The increased flow of air over skin surfaces has increased convective heat loss rates. Because relative humidity is low, it has also facilitated water vapor removal from the skin surface and promoted evaporative cooling. The net effect of just circulating more air in the office has not only made the environment more comfortable, it has also decreased the need to sweat as much. The C.E.O. was so pleased, he gave George a ride in his Mercedes.

Sample Problem No. 2 (Table 7-4, next page) presents circumstances of more severe heat stress. George's wife, Sally, runs a restaurant in an older section of town. The fifty year old building she uses has a kitchen that consistently is a problem. After George's success with the heat stress problem in his office, he and Sally use their borrowed test equipment to make environmental measurements in the kitchen as the first step in seeing what they can do about Sally's problem.

There are several factors that contribute to the discomfort and possible danger of working in the kitchen. Poor ventilation and air circulation, along with heat and water vapor produced by pots of boiling water and cooking food add to the problem. George and Sally estimate that most of the kitchen employees are working at about 700 BTU/hr, considering the different jobs they do. In evaluation of the "Before" information they've compiled, they decide that larger exhaust fans will help reduce humidity and improve air flow in the work area.

They're right. "After" measurements show that the "Heat Stress Index" has been markedly reduced, air flow is increased, and relative humidity is down. It's still a heat-stressing environment in which to work, especially as hard as is required of most kitchen employees, but the circumstance is improved. A major factor in making this a more comfortable and safe place is that with air temperature lower, there is a steeper thermal gradient for convective heat loss. Also, with relative humidity down, the water vapor partial pressure gradient is steeper which aides sweat evaporation and heat loss.

Chapter 7 - A Practical Problem

Table 7-4: Sample Problem No. 2 - A Hot Kitchen

Measurement	Before		After	
	English	Metric	English	Metric
Dry Bulb Temp.[1]	88.0	31.1	85.0	29.4
Wet Bulb Temp.[1]	75.0	23.9	60.0	15.6
Globe Temp.[1]	91.0	32.8	88.0	31.1
Air Velocity[2]	50.0	.25	400	2.0
Work Rate[3]	700	205	700	205
Calculations				
HSI	107.6	109.0	19.8	19.9
Exposure Time[4]	5.34	4.46	undefined	undefined
Radiation[3]	-18.6	-5.1	12.0	3.8
% Radiation	-2.9	-2.8	2.5	2.7
Convection[3]	-47.6	-13.8	-236.7	-69.2
% Convection	-7.5	-7.4	-49.8	-49.6
% Metabolism[3]	110.4	110.2	147.8	146.8
% Rel. Humidity	54.8	54.9	20.2	20.5
Evap. Required[3]	633.8	186.0	475.3	139.6
Evap. Maximum[3]	589.0	170.7	2400	702.9

Units: Note	English	Metric	
1	deg. F	deg. C	Slight differences between corresponding values are because of rounding. Calculations are for a hypothetical case only and are not intended to be applied in any real-life situation.
2	ft/min	m/sec	
3	BTU/hr	Watts	
4	Hr. min	Hr. min	

Chapter 7 - A Practical Problem

Also, raising air flow increases mass and energy transfer coefficients to facilitate both water and heat losses at skin surfaces.

G. Program Listing

Program HSI:

```
< < 1 3
 FOR CLEAR CLLCD
 0.2 WAIT
 "* HEAT STRESS INDEX * "
 1 DISP
 " Enter data ONLY in
  designated units"
 3 DISP
 "(any key to continue)"
 6 DISP TN
  NEXT 0 WAIT CLEAR
 1 CF TN
 "English or Metric
 units?(key E or M
 then press ENTER)"
 "" INPUT
  IF "M" SAME
  THEN 1 SF
  END TN
 "Dry Bulb Temp?" TD
 I DUP
  IF 1 FS?
  THEN CFC
  END 'TDB' STO TN
 "Wet Bulb Temp?" TD
 I DUP
  IF 1 FS?
  THEN CFC
  END 'TWB' STO TN
  IF 'TWB>TDB'
  THEN 1 3
   FOR CLEAR CLLCD
  0.2 WAIT
  "Wet bulb temp cannot
  exceed dry bulb temp"
  3 DISP
  "   (please wait)"
  5 DISP TN TN
   NEXT 3 WAIT HSI
   END '.555*(TDB-32
)'→NUM 'TDBC' STO
'.555*(TWB-32)'→NUM
'TWBC' STO
"Globe Temp?" TD I
DUP
 IF 1 FS?
 THEN CFC
 END 'TGT' STO TN
 IF 1 FS?
 THEN
"Air Velocity?(m/sec)"
"" I 'V' STO
 IF 'V<.1'
 THEN .1 'V' STO
 END 'V*196.8' →NUM 'V'
STO
ELSE
"Air Velocity?(ft/min)"
"" I 'V' STO
 IF 'V<20'
 THEN 20 'V' STO
 END
 END TN
 IF 1 FS?
 THEN
"Work Rate?(Watts)"
"" I 3.414 * 'W'
STO
ELSE
"Work Rate(BTU/hr)"
"" I 'W' STO
 END CALC > >
----------
```

Program CALC:

```
< <'(TGT-TDB)*V^.5*
0.13+TGT' UM 'MRT'
STO '(MRT-95)*15'
→NUM 'RD' STO '(TDB
-95)*V^.6*.65'→NUM
```

```
'CV' STO 'W+RD+CV'
→NUM 'EREQ' STO '
1000*ALOG(28.59-
8.2*LOG(TWBC+273
)+.00248*(TWBC+
273)-3142/(
TWBC+273))'→NUM
'PWB' STO '1000*
ALOG(28.59-8.2*
LOG(TDBC+273)+
0.00248*(TDBC+273
)-3142.3/(TDBC+
273))'→NUM
'PDB' STO 'PWB-
0.6748*(TDBC-TWBC)
'→NUM 'PH2O' STO '
PH2O*.75'→NUM 'PV'
STO '2.4*V^.6*(42-
PV)' →NUM 'EMAX'
STO
  IF 'EMAX>2400'
  THEN 2400 'EMAX'
STO
  END '100*(EREQ/
EMAX)'→NUM 'HS'
STO '250/(EREQ-EMAX
)' →NUM →HMS 'XPT'
STO
  IF 'XPT≤0'
  THEN "undef"
'XPT' STO
  END '100*(PH2O/
PDB)'→NUM 'RH' STO
TN TN CLLCD 1 FIX
HS "HSI= " SWAP + 2
DISP 2 FIX XPT
"XPT(hr:min) = "
SWAP + 4 DISP A
  IF 1 FS?
  THEN 'RD/3.414'
→NUM 'RD' STO
  END
  IF 1 FS?
  THEN 'EREQ/3.414
' →NUM 'EREQ' STO
  END IF 1 FS?
  THEN 'EMAX/3.414
' →NUM 'EMAX' STO
```

```
END CLLCD 1 FIX RD "RD= "
SWAP + DU
'100*(RD/EREQ)'
→NUM TN TN "%RD= "
SWAP + 4 DISP A
CLLCD TN TN
  IF 1 FS?
  THEN 'CV/3.4144'
→NUM 'CV' STO
END CLLCD CV
"CV= " SWAP + DU '
100*(CV/EREQ)'→NUM
"%CV= " SWAP + 4
DISP A CLLCD
  IF 1 FS?
  THEN 'W/3.414'
→NUM 'W' STO
END '100*(W/EREQ)
' →NUM TN TN
"%MET= " SWAP + 2
DISP RH "RH= " SWAP
+ " %" + 4 DISP A
TN TN CLLCD EREQ
"EREQ= " SWAP + DU
A CLLCD TN TN EMAX
"EMAX=" SWAP + DU A
TN TN CLLCD TN TN
"ANOTHER CALC?(Y/N)"
"" INPUT
  IF "Y" SAME
  THEN HSI
  ELSE TN TN TN
CLLCD
"*HEAT STRESS INDEX*"
2 DISP
"  PROGRAM OVER"
4 DISP 2 WAIT
  END CLLCD CLEAR {
RH XPT HS EMAX PV
PH2O PDB PWB EREQ
CV RD MRT W V TGT
TWBC TDBC TWB TDB }
PURGE>>
```

Program TD:

```
<<IF 1 FS?
  THEN "(deg C)" +
```

```
" "
  ELSE "(deg F)" +
" "
  END > >
----------
```
Program CFC:

```
< < 1.8 * 32 + > >
----------
```
Program I:

```
< < INPUT OBJ→ > >
----------
```
Program TN:

```
< < 2500 .1 BEEP 1000
2 BEEP > >
----------
```

Program DU:

```
< <
  IF 1 FS?
  THEN " Watts" + 2
DISP
  ELSE " BTU/hr" +
2 DISP
  END
> >
----------
```

Program A:

```
< <
" (any key to continue)"
7 DISP 0 WAIT
> >
```

Note: An alternative to the time consuming process of keying this program by hand is to transfer it into an HP-48G from a 3.5" or 5.25" disk using an IBM-compatible tabletop computer and an HP82208A or HP82208C Serial Interface. Another option is to copy it by infrared transfer from an HP-48S or HP-48G in which it is already stored. Either of these choices is easy and saves a lot of time and effort. For help with these procedures, see Chapter 8 and read, "Mastering the HP-48G/GX".

Notes

Chapter 8

Procedures for File Transfer

- Between HP-48G's
- Between HP-48G's, HP'48S', and *vice versa*
- Saving HP-48G View Window displays
- From an HP-48G to a Tabletop Computer
- From a Tabletop Computer to an HP-48G
- Sample Problems

Notes

Chapter 8 - Procedures for File Transfer

The HP-48G provides considerable storage space in its RAM. Most people will find that it's large enough to keep in memory any number of applications designed for personal and professional use. It can handle with ease, for example, many large programs, store various customized menus for conversion of complex sets of units, and keep on hand numerous user-designed equations. All the programs and applications described in this book, for example, will certainly fit into the HP-48G's RAM with lots of room to spare. For most users, an off-the-shelf HP-48G will provide many years of uninterrupted service before anyone has to be concerned with encroaching on its generous RAM capacity.

Even so, it is only a matter of time before practiced users will want to store data, programs and other kinds of HP-48G files elsewhere. They may not need to do this because they're running out of RAM, but because these files have applications in other places. For example, they may want to use a favorite customized program in another HP-48G without taking the time and going to the trouble to re-key all its steps. Such a program may very well have valuable extended use not just in one additional machine, but perhaps in the many HP-48G's of colleagues who also need to make similar calculations.

This chapter describes how to transfer stored LOCAL VARIABLES ("files") and subdirectories from one HP-48G to another. It also shows how to share them between an HP-48G and an HP-48GX, and between those in the earlier HP-48S computers. Once a few basic operations are understood, this kind of transfer is fast, easy, accurate and very useful.

Also, information stored in the HP-48G's ROM and some stored in its RAM may be valuably transferred either to 3.5" floppy disks, or the hard drive of a desktop computer. This material might be in the form of graphs, HP-48G view window displays, interim results or answers from final calculations, customized menus, programs, or other similar files. Once in magnetic storage, files can be held either in a library and later down-loaded back into an HP-48G, or transferred to some other model in the HP-48 series. Also, they can be shared with others by direct or electronic mailing services. Belonging to a network of colleagues within which programs of common interest are shared in this way has many advantages.

Besides just transferring files from magnetic storage, or from the memory of a desktop computer, they can be used in many other ways. For example, they can be incorporated either directly, or through favorite word processing or data management programs into printed text.

That's how, of course, the prints of program listings and all the screen displays were produced for this book. They were first generated in an HP-48G, then, with no more than a couple of keystrokes, quickly transferred to an IBM-compatible computer. They were then stored either as "*.BMP" or "*.TIF" files on a floppy disk. Imagine the savings in time and effort for transferring the HSI program (see "Program Listing" on pages 123- 125) in this way, compared to retyping the original text. This chapter describes these basic procedures.

There are many different ways to transfer, manipulate and store HP-48G files, as there are many different ways to perform successfully other of the machine's operations. The ones outlined in this chapter show just one way to do it. Others are encouraged to discover alternative techniques that work best way for them. The ones described here, however, will get anyone started.

Transferring Files between HP-48G's

Copying one or more subdirectories or files in a subdirectory from one HP-48G to another is the simplest of file transfer procedures. The following instructions are equally appropriate for making file transfers from an HP-48G to an HP-48GX model, or *vice versa*. Both machines will be identified as an "HP-48G".

Chapter 8 - Procedures for File Transfer

These file transfer procedures involve two HP-48G machines. The one that contains the subdirectories and files to be transferred is referred to as the "source". The one to which they will be transferred is referred to as the "recipient". To illustrate this file transfer process, three subdirectories with three files each have been generated. As shown in Figure 8-1, subdirectory "SUB1" contains files "A1", "B1", "C1". "SUB2" contains files "A2", "B2", etc.

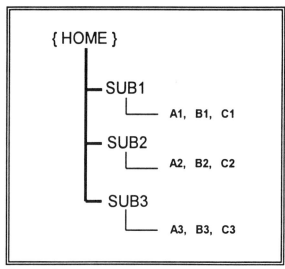

Figure 8-1

Generating the Subdirectories and Their Files

The "SUB1" subdirectory and LOCAL VARIABLES ("files") "A1", "A2" and "A3" are created in the "source" HP-48G by keying the following steps:

Step	Procedure
1	RS, MEMORY, NEW, ▼, α, α, S, U, B, 1, ENTER, ✔CHK, OK, CANCEL
2	SUB1, 1, ', α, A, 1, STO
3	2, ', α, B, 1, STO
4	3, ', α, C, 1, STO, RS, HOME

Create the subdirectories: "SUB2" and "SUB3" with their LOCAL VARIABLES in analogous ways. All three subdirectory titles will be listed in the view window display of the {HOME} directory (Figure 8-2).

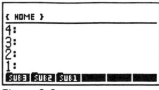

Figure 8-2

Transferring One or More Subdirectories

Problem Statement

Copy subdirectories "SUB1" and "SUB2" with their files from a "source" HP-48G to another HP-48G ("recipient").

Solution

Key the following steps:

Step	Procedure
1	Turn the "source" HP-48G on and go to the { HOME } directory (Figure 8-2).
2	Key: RS, I/O ("Send to HP 48..." is highlighted; Figure 8-3), OK (Figure 8-4), CHOOS (subdirectory "SUB3" is highlighted; Figure 8-5; next page), ▼, "✔CHK" (to select "SUB2" (Figure 8-6; next page).

Figure 8-3

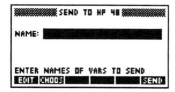
Figure 8-4

Chapter 8 - Procedures for File Transfer   151

Figure 8-5

Figure 8-6

3 Use the ▼ cursor control key to highlight subdirectory "SUB1", then key "✔CHK" to select it. A "✔" appears in the view window to the left of each selected subdirectory to indicate it will be included in the transferred group. When both subdirectories "SUB1" and "SUB2" have been identified (Figure 8-7), press "OK" (Figure 8-8).

Figure 8-7

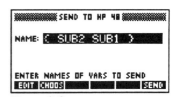
Figure 8-8

4 For the "recipient" HP 48G, key: RS, I/O, ▼ (to highlight "Get from HP 48...")

5 Place both the "source" and the "recipient" HP-48G's head-to-head, about an inch or so apart . Be sure the small triangle molded into the upper edge of one machine (just above the "4" in "48G" at the upper left of the case) is aligned with that of the other.

Chapter 8 - Procedures for File Transfer

6   First press "OK" on the "recipient" HP-48G, then "SEND" on the "source" HP-48G. First the message "CONNECTING" will be shown at the upper left of the view window, then a small arrow at the upper right of the view window flashes during the process of file transfer. File transfer is complete when the view window of the "recipient" HP-48G is once again clear.

7   Key CANCEL to clear the view window of the "source" HP-48G. "IOPAR" is removed by keying" ', "IOPAR", ENTER, LS, PURG.

Transferring one or More LOCAL VARIABLES

Problem Statement

Copy the LOCAL VARIABLE "C3" in subdirectory "SUB3" of a "source" HP-48G to the {HOME} directory of a "recipient" HP-48G.

Solution

Beginning at the {HOME} directory of the "source" HP-48G (Figure 8-2), key the following steps:

Step   Procedure

1   "SUB3", RS, I/O ("Send to HP 48..." is highlighted; Figure 8-3), ▼ (4 times to highlight "Transfer..."), OK (Figure 8-9), ▲, CHOOS, ▲ (to highlight "Infrared"), OK, ▼, CHOOS, OK (to select "Local vars..."), ✔CHK (to select "C:3"), OK (Figure 8-10)

Figure 8-9

Figure 8-10

2 For the "recipient" HP-48G, key: RS, I/O, ▼ (to highlight "Get from HP 48...")

3 Place both the "source" and the "recipient" HP-48G's head-to-head, about an inch or so apart . Be sure the small triangle molded into the upper edge of one machine (just above the "4" in "48G" at the upper left of the case) is aligned with that of the other.

4 First press "OK" on the "recipient" HP-48G, then "SEND" on the "source" HP-48G. First the message "CONNECTING" will be shown at the upper left of the view window, then a small arrow at the upper right of the view window flashes during the process of file transfer. File transfer is complete when the view window of the "recipient" HP-48G is once again clear.

Transfers Between the HP-48G and the HP-48S, and *vice versa*

There are many advantages to working even informally in a network of HP48 users. It's an excellent way to share computational skills and programming tricks. But, not everyone will have an HP-48G (or HP-48GX). Some will continue to use the older, but very similar and serviceable , HP-48S (or HP-48SX). The following sections describes how to transfer stored data between these machines. Instructions are equally appropriate for both models in the HP-48S series, and for both in the HP-48G series.

File Transfer from an HP-48S to an HP-48G

For purposes of illustration, the subdirectory structure shown in Figure 8-1 has been constructed in the HP-48S' {HOME} directory.

Problem Statement

Copy the subdirectory "SUB3" and its files from an HP-48S to the {HOME} directory of an HP-48G.

Key the following steps:

For the HP-48S:

Step	Procedure
1	LS, { }, "SUB3", ENTER
2	LS, I/O, SETUP, (if necessary, press white key "A" to toggle from selection (top of view window) of "wire" to IR (for "infrared")), ATTN
3	(With "{SUB3}" still displayed at bottom of view window), LS, I/O

For the HP-48G:

Step	Procedure
1	Go to {HOME} directory
2	RS, I/O, ▼ (to highlight "Get from HP 48…")
3	Place the "source" HP-48S and the "recipient" HP-48G's head-to-head, about an inch or so apart . Be sure the small triangle molded into the upper edge of the HP-48G (just above the "4" in "48G" at the upper left of the case) is aligned with that of the HP-48S (Just above the end of the word "Hewlett").
4	First press "OK" on the "recipient" HP-48G, then "SEND" on the "source" HP-48S.

File Transfer from an HP-48G to an HP-48S

For purposes of illustration, the subdirectory structure shown in Figure 8-1 has been constructed in the HP-48G's {HOME} directory.

Chapter 8 - Procedures for File Transfer   155

Problem Statement

Copy the LOCAL VARIABLE "B2" from the subdirectory "SUB2" in an HP-48G to the {HOME} directory of an HP-48S.

Key the following steps:

For the HP-48G:

Step	Procedure
1	(In the {HOME} directory), "SUB2", RS, I/O ("Send to HP 48..." highlighted), OK, CHOOS, highlight "B2: 2", ✔CHK, OK

For the HP-48S:

1	LS, I/O
2	Place the "source" HP-48G and the "recipient" HP-48S head-to-head, about an inch or so apart . Be sure the small triangle molded into the upper edge of the HP-48G is aligned with that of the HP-48S.
3	(For the HP-48G), SEND, (for the HP-48S), RECV

The remaining procedures described in this chapter involve connecting the HP-48G to a desktop computer. As when connecting any peripheral to a computer, be sure the desktop computer and the HP-48G are both turned off. The connection between the two machines is made with a serial interface cable. Two models work equally well: the HP82208A and the HP82208C.

The smaller end of the interface cable inserts into the RS232 connector at the upper edge of the HP-48G. Appropriate caution is necessary to match the geometry of the cable end (rounded side down) to that of the HP-48G and not to bend or damage the small pins recessed into its plastic case. The larger end of the interface cable inserts into the serial connector on the desktop computer.

Operations using the interface cable are activated and controlled by the software that comes with it. The software supports many more functions than just transferring files, as described in the "README.TXT" files on the disk that comes with each model of interface cable.

The first step in using the interface cable is to load the operating files from the supplied program disk to the "C:\" drive of the desktop computer. This can be done just as well directly from the DOS display, or through WINDOWS 3.1 or WINDOWS 95 displays. The basic procedure involves activating "SETUP" on the program disk. This copies files to the "C:\" drive of the desktop computer.

If the model HP82208A interface is used, operating files will be stored in a subdirectory called "SERIAL". They will be in a subdirectory called "LINK48", if the HP82208C model is used. Instructions that come with the interface cable and those in the program disk's 'README.TXT" file describe how to configure the desktop computer. An icon entitled, "HP48 Grabber" will appear on the operating display of the desktop computer when the interface cable software has been correctly installed.

Saving the HP-48G View Window Display to a Disk

Being able to use an HP-48G view window display in a printed document is of great utility. This is easily done when the HP-48G is connected by an interface cable to a desktop computer. The first step is to capture the display, then save it to a disk. The next step is recall the view window display file and, if necessary, change its overall size, add a border and a caption, and include it as graphic in the developing document. This is done easily with most word processing and graphics software programs.

It is necessary to set a FLAG -34 in the HP-48G for it to work correctly with the interface cable. This needs to be done only once, unless the FLAG is intentionally cleared later. Once set, it remains set, even when the calculator is turned off.

Key the following steps to set FLAG -34: 3, 4, +\-, PRG, TEST, NXT, NXT, SF.

Problem Statement

Construct and save as a file the "Figure 8-2" at the beginning of this chapter.

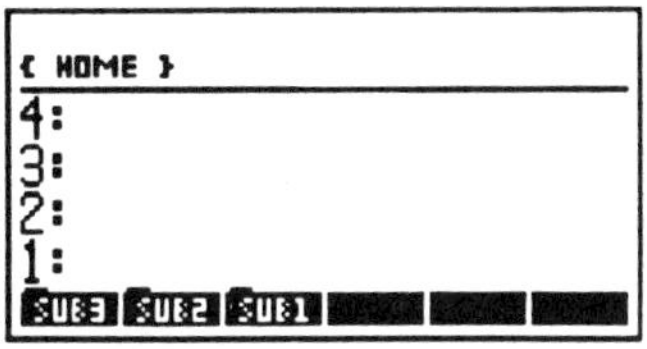

Figure 8-2

Solution

Key the following steps:

Step	Procedure
1	Turn off the desktop computer and the HP-48G, and connect the interface cable between them. Turn on the table top computer, wait for its operating screen to be displayed, then turn on the HP-48G.
2	Construct the subdirectories "SUB1", "SUB2" and "SUB3" in the {HOME} menu of the HP-48G.
3	Double-click on the "HP48 Grabber" icon shown on the desktop monitor.
4	(On the HP-48G) RS, I/O, ▼ (twice to highlight "Print display"), OK. (As an alternative, hold down the "ON" key and press "1")
5	Once the captured view window is in the "HP48 Grabber" display, click on SAVE, select the drive to which the file will be saved, designate the file format ("*.BMP" or "*.TIFF"), type a file name and save it.

File Transfer from an HP-48G to a Desktop Computer

There are at least two benefits from transferring data, programs and other stored information from the HP-48G to a desktop computer. They can then be incorporated directly into documents produced by the computer. They can also be transferred to storage on magnetic media like floppy disks. This is a useful technique for sending programs to others to be downloaded into their own HP-48G's, as described later in this chapter.

The following two procedures assume:

1. an HP-48G has subdirectories and files constructed as shown in Figures 8-1 and 8-2,

2. the HP-48G is connected by a serial interface cable to an IBM-compatible desktop computer,

3. both the HP-48G and desktop computer are turned on, are correctly configured to work with the interface cable, as described earlier in this chapter, and that

4. the KERMIT software supporting the interface cable is stored in drive "C:\" of the desktop computer.

Problem Statement

What is the procedure for transferring a subdirectory with its files, or individual files, from an HP-48G to a desktop computer?

Solution

Complete the following steps:

For the IBM-compatible desktop computer:

Step	Procedure
1	If the HP82208A interface cable is used, run "C:\SERIAL>KERMIT". If the HP82208C interface cable is used, run "C:\LINK48>KERMIT". (read: "Kermit-MS>")
2	Set port "by: "Kermit-MS>SET PORT 1"
3	Set baud rate by: "Kermit-MS>SET BAUD 9600"
4	Prepare for transmission by: "Kermit-MS>RECEIVE"

For the HP-48G:

Step	Procedure
1	Key: RS, I/O, ▼ (4 times to highlight "Transfer..."), OK
2	If necessary, configure display to: "PORT: Wire", "TYPE: Kermit", "FMT: ASC", "XLAT: New1", "CHK: 3", "BAUD: 9600" and "PARITY: None".
3	Use cursor control keys to highlight "NAME:", then press CHOOS, (highlight "Local vars"), OK.
4	Highlight each directory or file to be transferred and press "✔CHK" to mark each with a "✔".
5	Press "SEND" and wait for file transfer to be completed.

Transferred files are now copied with the same name either to the subdirectory "C:\SERIAL" or to the subdirectory "C:\LINK48". Each can now be edited, moved and copied to other subdirectories or imported for use either into word processing, or other programs. When moving or copying transferred files, it is important that other files in "C:\SERIAL" or "C:\LINK48" remain intact.

Chapter 8 - Procedures for File Transfer

Problem Statement

What is the procedure for copying a file from storage in a subdirectory in a desktop computer to an HP-48G?

Solution

The following procedures support transferring one file at time. (Hint: use all capital letters for file names)

For the IBM-compatible computer:

Step	Procedure
1	If the HP82208A interface cable is used, run "C:\SERIAL>KERMIT". If the HP82208C interface cable is used, run "C:\LINK48>KERMIT". (read: "Kermit-MS>")
2	Set port "by: "Kermit-MS>SET PORT 1"
3	Set baud rate by: "Kermit-MS>SET BAUD 9600"
4	Prepare for transmission by: "Kermit-MS>SEND", ENTER (read: "Local Source file:"), type name of file in either the "C:\SERIAL", or the "C:\LINK48" subdirectory, press ENTER (read: "Remote Destination File"), type name of file to be transferred, press ENTER.

For the HP-48G:

Step	Procedure
1	Construct and enter the name of the subdirectory into which files will be transferred

Chapter 8 - Procedures for File Transfer

2   Key: RS, I/O, ▼ (4 times to highlight "Transfer..."), OK

3   If necessary, configure display to: "PORT: Wire", "TYPE: Kermit", "FMT: ASC", "XLAT: New1", "CHK: 3", "BAUD: 9600" and "PARITY: None".

4   Press "RECV" and wait for file transfer to be completed.

Notes

Index

Notes

Index

A

ALPHA 3, 89, 92, 93, 116, 117
ALPHA statements 93
Animals Regulate Body
    Temperature (How) 122
apostrophe key 13, 15, 30, 94,
    118, 119
applications 1, 2, 10, 12, 13,
    16, 65, 72,117, 121, 147
arithmetic operations 3, 10, 67
Arithmetic Operations Using
    Matrices 67
ARRAY (array) functions 106
average 43, 74, 102, 111, 115,
    116,126, 129

B

basic arithmetic operations 10
basic calculations 117
basic directory structure 12
basic operations 2, 7, 147
basic statistics 101, 113
Basics of Program Structure 83
Beam Deflection 41
blood flow 125, 126
body heat content 122, 123,
    131
body heat gain 125
body heat loss 126
body surface 74, 126-129
body temperature 122-126,
    129, 133
Brush Turkey 73

C

Calculator Modes 9, 10
camel 128
CANCEL 7, 15, 38, 60, 65, 76,
    84, 85, 88, 101, 106, 118,
    149, 152
CANCL 33, 84
Capacitive Energy 43
CHOOS 31, 76, 77, 150, 152,
    155, 159
CLLCD 91, 94, 113, 116, 141,
    142
Column Loading 38
complex numbers 65, 72
conditional test 117
conditional tests 83
conduction 75, 124, 129
conductive 121
Construct a Matrix 71
constructed display 92
control system 122, 123
convective heat transfer
    125-127, 133, 135
correlation coefficient 111, 112
covariance 111, 112
CST key 9
curve-fitting 106, 119

D

data arrays 101
data be cleared 108
data storage 11, 12

decimal 9, 10, 70, 76, 92, 102, 104, 113, 115, 116
degrees 50, 57
deleting LOCAL VARIABLES 117
DEMO 21, 26, 28, 86, 91, 94, 95
dependent variable 108, 111
Designing the Program 89
desktop computer i, 1, 147, 148, 155-159
DET 67, 69
Determinant of a Matrix 67
differential equation 60, 61
DIR 119
Directory Structure 12
Drag Force 46, 47

E

ECHO 85
EDIT 33, 59-62, 67, 68, 95, 101
Editing a Matrix 67
EGVL 67, 70
eigenvalues 67, 70
Electrical Resistance 42
emissivity 127
energy 43, 125, 127, 140
engineering applications 72
entering data 113
environmental heat and cold exposure 123
enzyme 106-108, 112
EQN 24, 39, 41, 42, 44
EQUATION LIBRARY 1, 2, 12, 14, 16, 19-21, 27, 28, 85
Equation Writer 1, 2, 12, 19, 27-29, 33, 39, 88

evaporation 75, 123, 128-130, 139
evaporative 121, 129, 132, 133, 135, 139
evaporative heat transfer 121
exercise 13, 14, 27, 37, 38, 41-61, 108, 123, 125, 126, 129

F - G

file transfer 147-149, 152-154, 158-160
first-order differential equation 60
fish 122
FIX 9, 72, 91, 94, 102, 113, 115, 116, 142
flow diagram 85, 133
Fluid Dynamics 43
Fluid Pressure 44
Gas Expansion 47
Generate the Matrix 67
generating ALPHA 93
geometry 53, 54, 73, 155
Getting out of Traps 7
globe temperature 133, 135
graphical solution 76

H

heat 21, 22, 28, 31, 33, 48, 73-78, 86, 87, 89, 91, 121-133, 135, 136, 138, 139, 141, 142
heat exchange 123-127, 131, 135

Heat Exchange by Convection 125
Heat Exchange by Thermal Radiation 127
heat flow 21, 28, 89
heat gain 123, 125, 128-130
Heat Loss by Evaporation 129
heat production 73-76, 121, 123, 128, 130-132, 135, 136
heat strain 131, 132
heat stress 121, 122, 129, 131, 132, 135, 136, 139, 141, 142
Heat Stress Index 132, 135, 141, 142
heat transfer 48, 75, 121, 124-128, 133, 135
heat transfer coefficient 126
HOME directory 12-15, 21, 101, 117, 118
HOME menu 10
HOME stats 103, 106, 108, 111, 112
hot kitchen 140
hot office 133, 137
HP82208A; HP82208C 143, 155, 156, 158, 160
HP-48G keyboard 2
HSI 121, 122, 132, 133, 135, 137, 140-142, 148
huddling 128

I - J

IBM-compatible 143, 148, 158, 160
Importance of subdirectories 14

independent variable 108, 111
infrared transfer 143
input 45, 72, 73, 83, 86, 89, 91, 92, 95, 113, 141-143
insects 122
interface cable 1, 155-160
internal temperature 122
inverse of a matrix 67
IR 154
Isothermal Gas Expansion 47
iteration 76

K - L

KERMIT 158
kitchen 139, 140
Left-shifted Function 27
linear equations 61, 65, 70, 73, 79
Linear Motion 51
local variable 13, 27, 30, 40, 86, 92, 96, 103, 116, 118, 152, 155
LOCAL VARIABLES 1, 14, 24, 26-28, 33, 40, 42-49, 51-57, 70, 91, 92, 95, 96, 101, 106, 117-119, 147, 149, 150, 152
LS, CLEAR 11, 14, 33, 40, 85, 88, 94, 106, 111
LS, SWAP 11

M

Mass-spring Oscillation 53
MATR 67, 69, 70
matrices 67, 68

matrix 62, 65-68, 71, 101-104, 106
matrix operations 65, 68
matrix table 67, 101-104, 106
mean 106, 113, 116, 117, 135
measured data 107
Metabolic Heat Production 121, 128, 130-132, 135, 136
mistakes 95
model 39, 41, 42, 44-46, 49, 50, 54-57, 107, 112, 148, 156
Mohr's Circle 57
motion 50, 51
MTH 67, 69, 70, 73

N - O

NMOS transistor 55
NORM 67, 69
numbers are displayed (How) 9
NXEQ 24, 39
NXT 12, 13, 25, 26, 39-42, 44-46, 49, 51, 56, 57, 67, 69, 70, 84, 94, 95, 108, 118, 157
office 121, 133, 136-139
OK 15, 31, 38, 58-60, 62, 65, 76, 77, 101, 111, 149, 150, 152, 155, 157, 159, 160
oscillation 53
output 83, 86, 89, 92, 125

P

PGDIR 119

physical activity 130, 131
Physical Avenues of Heat Exchange 124, 131
PIC 23, 24, 39, 41, 42, 44-46, 49, 50, 54-57
placing data 106
Plane Geometry 53
POLAR 53, 56
polynomial 58
predict 19, 112, 131
primary function 7, 129
processing 83, 86, 89, 92, 148, 156, 159
Program Construction 97
Program Description 91, 115
Program Design 85
Program Listing 91, 113, 133, 141
Projectile Motion 50
PURG 14, 26, 33, 40, 68, 94, 106, 118, 152
purge 12, 15, 27, 28, 42-49, 51-57, 91, 113, 142

Q - R

quadratic equation 72
radiant exchange 127
radiant heat 127, 128, 131, 138
radiation 127-129, 133, 137, 140
radiative 121, 127, 128, 132, 135
RAM 1, 101, 108, 147
reciprocal 9, 70
repetitive operations 91

reptiles 122
RESET 108
Reverse Polish Notation 10
Reynolds Number 43
Right Cursor 94
Right-shifted Function 27
ROM 20, 83, 102, 116, 147
Roots of a Polynomial 58
RPN 3, 10, 11
RS, MATRIX 62, 66, 68, 71
RS, MEMORY 38, 65, 101, 118, 149
RS, STAT 103, 108, 111, 112

S

sample problem 20, 65, 85, 136, 137, 139, 140
Saving the HP-48G View Window Display 156
screen remains blank 7
SDEV 113, 116
second-order differential equation 61
self-regulating temperature 74
serial interface 143, 155, 158
Shear Stress 56, 57
Simultaneous Solutions for Non-linear Equations 73
single data set 102
Solid Geometry 54
Solid State Devices 55
SOLV 24, 40-56, 58, 59
solve functions 27
solved problems 37
Solving a System of Linear Equations 70
sound a tone 92

source 2, 130, 131, 160
SPC 58, 59, 62, 66, 68, 71-73, 85, 94, 95
special applications 65
special characters 12, 83, 84
stable internal temperature 122

stack 2, 8-14, 29, 33, 65, 92, 93, 106, 108, 111, 115, 116
stack designations 8
standard deviation 101, 102, 107, 111, 115-117
standard notation 9

statistical calculations 112
statistical operations 101, 106
statistical table 101, 108, 111, 115, 116
statistics i, 101, 102, 105, 106, 113, 119
STD 72, 110, 111
STO 13, 30, 62, 67, 68, 76, 77, 88, 91, 94, 95, 113, 141, 142, 149
stress 56, 57, 121, 122, 124, 129-133, 135, 136, 138, 139, 141, 142
subdirectories 2, 12-14, 37, 83, 91, 117, 118, 147-151, 157-159
subdirectory 14-16, 21, 26-28, 31, 33, 37, 38, 40, 65, 76, 83, 86-88, 91, 94, 95, 101, 106, 108, 115, 118, 119, 148-156, 158-160
substrate concentrations 107, 108, 112

sweat 123, 129, 130, 139
sweat glands 129, 130
symbol definitions 135
Symbolic Solution 86, 87
System of Linear Equations 61, 65, 70

T - U - V

temperature 21, 28, 48, 49, 73-76, 102, 122-131, 133, 135, 138, 139
Terminal Velocity 52
thermal balance 124
thermal conduction 75, 124, 129
thermal conductivity 21, 28, 85, 86
thermal radiation 127, 128, 133
tissue damage 122
tone 10, 89, 91, 92, 94-96, 115-116
total 102, 113, 115-117, 125, 128, 129, 133
transcendental functions 8
transistor 55
user-constructed variable menu 13
user-defined name 27
using equations 19
Using the Equation Library 12, 20, 27, 28
using the program 89, 133
variance 102
vectors 65, 72
view window display 150, 156

W - X - Y - Z

WAIT 91, 94, 113, 141-143, 157, 159, 160
water 43, 46, 47, 52, 73, 126, 128, 129, 131, 135, 139, 141
white-faced keys 3
Wind Force 45
writing programs 1, 83, 93
X STACK REGISTER 8-10, 13, 29, 108, 116
Y STACK REGISTER 8, 10, 108, 116
zero is entered 116
0 (zero) FIX 113, 116
0 (zero) WAIT 113, 141, 143